Biofuel's Engineering Process Technology

Biofuel's Engineering Process Technology

Editor

Sabbas Radley

scitus academics

Biofuel's Engineering Process Technology
Edited by **Sabbas Radley**

ISBN: 978-1-68117-225-5
Library of Congress Control Number: 2016934760

www.scitusacademics.com
Box No. 4766, 616 Corporate Way,
Suite 2, Valley Cottage,
NY 10989

Notice

Printed in United States of America on Acid Free Paper

Preface

Bioenergy is energy produced from biofuels. Overall, bioenergy accounts for about 10% of the total energy demand in the world. Traditional unprocessed biomass such as fuel wood, charcoal and animal dung is the main energy source for many people in developing countries who use it primarily for cooking and heating purposes. More advanced and efficient conversion technologies now allow the extraction of biofuels from materials like wood, crops and waste. Biofuels can be solid, gaseous or liquid, although often in the literature only liquid biofuels for transport in a narrow sense are referred to. Biofuels can be derived from agricultural crops, including conventional plants for food and power. Biofuels may also be derived from forestry, agricultural or fishery products or municipal waste, as well as agro-industry, food industry and food service by-products and waste products. Bioethanol and biodiesel are currently two major biofuels. Bioethanol is used as a petrol substitute and biodiesel is used as a diesel substitute. Biofuels are an enormous growth opportunity worldwide and have an important role to play in replacing fossil fuels worldwide with a cleaner alternative for renewable sources.

Highlighting the concepts, systems and technologies involved in plant and animal feedstock fuel production, this book is intended to summarize the concerns of current biofuels and thus to contribute to this important topic. Biofuels are also expected to continue in the foreseeable future and everyone interested in this varied and evolving subject will be able to read the biofuels features covered in this book.

Table of Contents

1

CHAPTER

NANOTECH BIOFUELS AND FUEL ADDITIVES

Sergio C. Trindade

[1]*SE2T International, Ltd. and International Fuel Technology, Inc., USA*

1. INTRODUCTION

This chapter was inspired by an invited presentation of the author at the Chemindix conference in Bahrain in October 2010. This was the 8th. International Conference & Exhibition on Chemistry in Industry, promoted by the Saudi Chapter of the American Chemical Society and Aramco. The focus is on reviewing the application of nanotechnology to biofuels production and to the utilization of fuel additives, some of which are derived from renewable materials.

To introduce the topic, the broader context of petroleum fuels and biofuels is presented. A smart future of oil refining would be to increasingly utilize margins to finance a transition away from oil towards future alternative providers of mobility, in particular biofuels.

Future scenarios of liquid biofuels involve the market penetration of second and further generations of technologies and the continuous improvement of first generation processes. On the other hand, nanotechnologies are among the candidate technologies for the biofuels of the future. The nanotechnology field is vast and its applications unbound.

This is followed by a brief review of nanotechnology developments, especially as they apply to liquid particles, beyond the more common solid particle applications.

Algae growth, harvesting and conversion are presented and discussed, given the immense potential of their contribution towards an energy future where biofuels play a significant role.

Most of the current effort in second generation conversion to liquid biofuels is based on biomass cellulosics to ethanol and biodiesel. Nano processes are being pursued and will be reviewed in the chapter.

Likewise, the presently used processes to convert oils and animal fat into biodiesel are based on trans-esterification with methanol or ethanol, which inevitably generates glycerol, which must find a market or get disposed properly. Nano processes may be useful in addressing this issue.

Speculative considerations are made about the role of liquid nanoparticles of fuel additives in enhancing the performance of additized biofuel/fuel blends, in connection with surface and combustion effects.

Public concerns over the impacts of nanotechnologies on security, health and the environment are also mentioned and discussed. But, a cautionary optimistic view is presented on the huge benefits of a careful penetration of nanotechnologies in the realm of biofuels and fuel additives, and in many more applications, especially those dealing with human health.

2. THE FUTURE OF OIL REFINING [1] AND THE OIL TRANSITION TO ALTERNATIVES [2]

Crucial challenges to the oil industry are evolving, as the demand for energy services (mobility, lighting, rotary movement, heating and cooling) increases, which with the current technology setup, is translated into expanding demand for liquid fossil fuels. Oil producers and refiners face difficulties finding sufficient good quality crude oil in adequate amounts and reasonable costs to meet growing demand over the long run, while users and the public at large are pushing for environmental improvements, such as better air quality in the immediate future. Moreover, concerns about the impacts of climate change caused by increased greenhouse gases emissions from the production and use of liquid hydrocarbons may eventually force a transition to climate friendly energy services providing systems. This offers biofuels a market penetration opportunity in the transition towards a yet undefined new energy future.

Under this background, the oil refining industry in the USA and the European Union has been stagnant. It has been immobilized by environmental obstacles posed by an articulated public, augmented by a "not in my backyard" attitude that makes it difficult to build new refineries. In addition, declining margins for refined products may have led major players to focus more on the upstream.

On the other hand, refining is expanding in other parts of the world, such as India, China, Brazil and the Middle East, as these countries develop and the oil producers attempt to add more value to their resources. An evidence of the shift of refining towards developing economies is the fact that the largest refinery in the world is in India and belongs to Reliance.

But, all over the world, the increase in the long-term marginal cost of oil combined with environmental pressures and stricter government regulations and mandates are likely to lead to the decline of the centrality of oil in the global energy mix in favor of natural gas. This shift in dominance happened to wood and coal over the past two centuries and is now happening to oil. Oil companies are increasingly calling themselves energy companies. Some of them will leverage their current oil production margins to make a smooth transition to alternatives over time. A profitable transition to alternatives in the oil economy would require a gradual transfer of oil profits into green investments and the stretching of current oil supplies.

3. LIQUID BIOFUELS ISSUES [3]

The enormous global daily consumption of liquid fuels is of the order of 80 million barrels/day (e equivalent of 12.7 million m^3/day). The sugar cane area required to produce the same volume of ethanol is about 700 million hectares, assuming a yield of 6.5 m^3/ha/year of ethanol. This area is equivalent to 100 times the sugar cane cultivated area in Brazil, the second largest bio-ethanol producer in the world. Biofuels definitely face an issue of scale. In 2010, fuel ethanol and biodiesel combined displaced a mere 3% of oil in the world.

Figure 1[4] below illustrates the scale issue by showing how much land it would take for the USA to grow its own fuel.

It appears that algae require the least area to meet the large scale demand of liquid fuels in the USA, whereas the area required by soybeans is larger than the USA's 48 continental states. The area required by corn is substantial. This suggests that the current biofuels production base of the USA would not be able to meet demand, and imports would be required to meet the colossal American energy appetite.

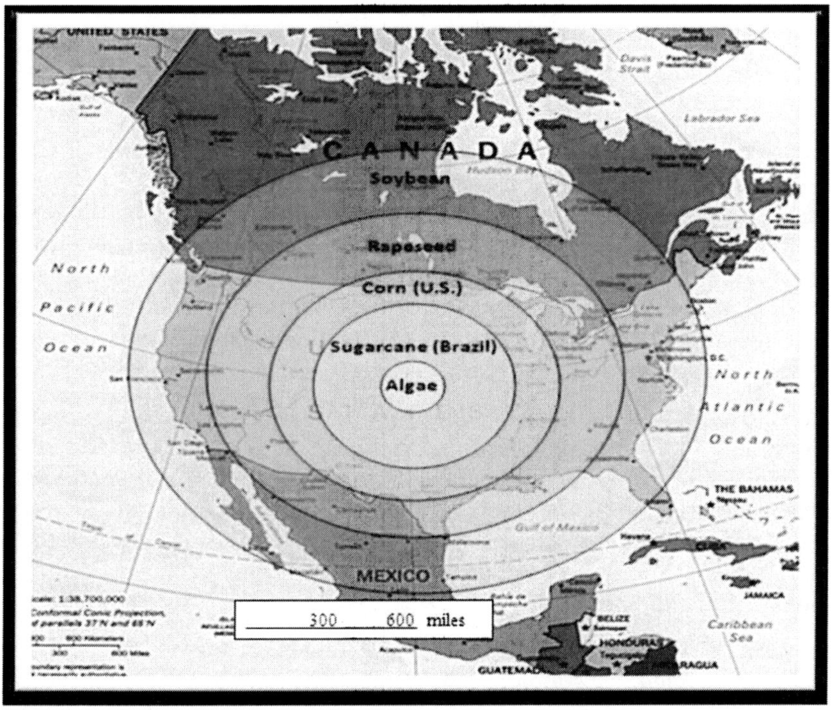

FIGURE 1. How much land would it take for the USA to grow its own fuel?

The scale challenge posed to biofuels relates to the labour, management, land, water and sunshine required to produce the biomass and the processing that originates them. These are scarce resources that are also needed to grow food, feed and fibre to ultimately meet various human demands. These are resources that have an opportunity cost from competing markets. To develop biofuels in the scale of commercial liquid fuels require massive financing, a resource that may have alternative uses as well. The mobilization of private capital under a perception of market and other uncertainties is another issue that biofuels have to resolve in order to thrive.

The production of biofuels is accompanied by local environmental issues that need addressing. For instance, in the case of sugar cane ethanol, stillage the liquid residue of distillation, has a high chemical and biological oxygen demand and requires appropriate processing before final disposal. From a global climate change perspective, designed and managed properly, a biofuels production system would add minimally to greenhouse gas emissions. But, in practice, many biofuel production systems in the world are contributing net GHG emissions.

A bone of contention in the development of the biofuels industry is the present competition for feedstocks between the food and fuel industries. In the case of biodiesel, all commercial vegetable oils that are used in preparing food are also convertible to biodiesel. A similar situation exists with respect to fuel ethanol, especially for the starch-based feedstocks (corn and wheat). However, the hike in food prices that happened globally in 2007/8 and is happening in 2010/1 derive mostly from other causes such as droughts and other climate related phenomena, higher oil prices and market speculation.

Since the cost of biofuels is dominated by feedstocks cost, access to feedstocks in the required amounts, timing and at adequate prices is key to the success of the biofuels economy. The combination of the food versus fuel conundrum with the need to have reliable and economic access to feedstocks is shifting the industry towards non-food feedstocks and to the market penetration of second generation technologies to convert cellulosic biomass into liquid biofuels.

Concern in important consuming markets about the sustainability of biofuels producing systems is putting pressure on suppliers to abide by sustainability protocols subject to certification. The sustainability of biofuels is actually linked to freer international trade, which would tend to phase out unsustainably produced biofuels in favour of regions of the world that can meet sustainable production requirements. A valuable discussion on this matter was hosted by the Rockefeller Foundation in 2008 at its Bellagio Centre and produced a sustainable biofuels consensus. The objective was to understand the many drivers for sustainable trade, consumption and production of biofuels, and the comparative advantage of supplying regions combined with demand and technology from consuming regions [5].

However, much remains to be done to achieve free international trade of biofuels. The World Trade Organization Doha rounds have reached an impasse. Currently, biodiesel is considered an industrial product, whereas fuel ethanol is categorized as an agricultural product, which allows more protectionism. What is needed is a unified treatment of biofuels, where they are classified under environmental goods and services. But, irrespective of these drawbacks, a sign pointing to a larger role for biofuels in the future are the new biofuels technology initiatives by large oil companies, such as BP, Chevron, Exxon and Shell.

The development of the international trade in biofuels is likely to distribute more evenly the production and consumption of biofuels in the world. For the time being, biofuels production is overwhelmingly concentrated in the USA, Brazil and the European Union, as shown in Fig. 2 below [6].

4. THE VASTNESS OF NANOTECHNOLOGY

Nanotechnology can be simply defined as the discipline of building machines/devices on the scale of molecules, a few Nanometers (10^{-9}m) wide, way smaller than a cell. Table 1 below show some practical applications of nanotechnologies and confirms the vastness of their domain [7]:

In the practically important area of polymers, nanotechnologies originate nano-structured polymers, where applications can be found in support structures; manufacturing processes; diagnostics and therapy; pharmaceuticals; medical and dental prosthesis; and thin films for surface treatment. The main chemicals involved in nano-structured polymers are: poly-oxides; poly-acrylates; poly-vinylics; poly-saccharides; and poly-ethylenes. The main materials incorporated into polymer nano-matrices are silicon, chromium and carbon [8]

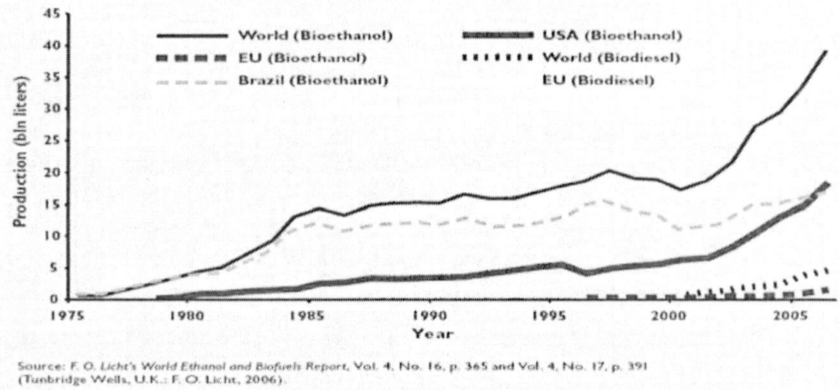

FIGURE 2. World's biofuels production is concentrated in the US, Brazil and the EU

Table 1. Some applications of nanotechnology

Nanocrystalline Drugs	Oxide Nanoparticles
Nanofilm Coatings	Carbon Nanotubes
Cosmetics	Silicon Nanomaterials
Nanocatalysts	Metal nanoparticles
Ceramics	Polymer Nanocomposites
Rubbers	Inorganic Nanocrystals

5. TURNING ALGAE INTO BIOFUELS

As shown in Fig. 1, biofuels derived from algae offer a great potential in view of the possible high yields and smaller area requirements. In addition, algae can play a role in carbon mitigation, as one way of growing algae is to feed them carbon-dioxide (CO_2), besides water and sunlight. Algae can be fed other substrates as well, because to grow, cost-effectively, on carbon dioxide there would be a need of concentrated sources of the gas, such as found in combustion off-gases from fossil fueled power plants.

Oil can form up to 50% of the algae mass, in contrast with the best oil-bearing plants – oil palm trees – where less than 20% of the biomass is made out of foil. Algae carbohydrates can also be made into ethanol or gasified into bio-gas, or methane or hydrogen [9].

But, algae development into biofuels must overcome a number of challenges before algae can become significant sources of commercial biofuels. Since algae also need water to grow, expansion of algae production may create a dilemma of water versus fuel, similar to food versus fuel dilemma discussed previously. Another challenge is the low natural carbon dioxide concentration in the atmosphere, hence the consideration of additional sources of carbon for algal growth in a commercial biofuels system. One response to these challenges may include the use of nanotechnology to turn algae into biofuels.

As way of examples, in 2009, the company Quantum Sphere received a grant from the California Energy Commission to develop a nano-catalyzed algae biogasification. Also in California, the Salton Sea receives large amounts of agricultural runoff, which sometimes create large algae blooms. These algae and similar biomass have been turned experimentally into methane, hydrogen and other gases [10].

One nanotechnology relevant to algae development is the use of nanoparticles as no-harm harvesters of biofuel oils from algae, as illustrated in Fig. 3 [11]. The nano particles are shown on the left hand side of the photograph before the oil pregnant algae are added. The right hand side shows the contacting between the algae and the nano particles, which results in extracting the oil without harming the algae. Maintaining the algae alive can dramatically reduce production costs and the generation cycle.

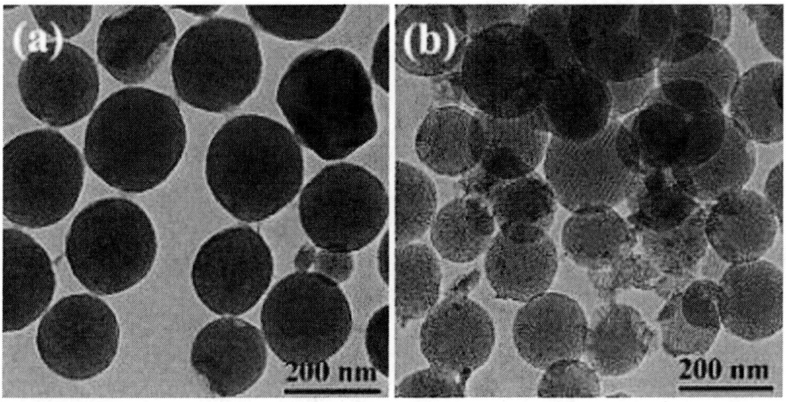

FIGURE 3. Nano-particles harvesting oil from algae without harming the organism

One possible downside of the nano-harvesters is the risk that they may be released into the environment, although the spherical nano-particles are made of calcium compounds and sand [12]. The pores of the spheres are lined with chemicals, which extract algal oil without breaking the cell membrane. Nevertheless, prior to commercial market penetration of nano-harvesting, there would be a need to carry out due diligence to ensure the safety of these processes.

6. NANOTECHNOLOGY APPLIED TO LANDFILL FACILITIES[13]

The organic matter in landfills tend to undergo anaerobic fermentation yielding methane and CO_2 [14], which if naturally vented into the atmosphere would add to the greenhouse emissions that warm the climate. And the climate change impact of methane is 25 times larger than that of carbon dioxide for a time horizon of 100 years [15]. Thus, there is a need to sequester the carbon present in landfill methane. Nano-catalysts can crack methane into elemental carbon and hydrogen. The carbon can be produced in high-purity nano-graphite for use in aerospace, automobile, batteries, etc. This approach to handling methane can considerably improve the economics of landfills as well as of anaerobic digester plants that generate electricity from biogas fueled electricity.

7. NANOTECHNOLOGY TO CONVERT BIOMASS INTO BIOFUELS

Delinking biofuels production from food crops is a necessary condition to expand the scale of the market penetration of biofuels globally. Among the challenges this strategy faces is the inherent resistance of cellulosic feedstocks to conversion to simpler sugars that can be fermented into ethanol. Here, the promise lies in nano-particles used as immobilizing beds for expensive enzymes that can be used over and over again to breakdown the long chain cellulose polymers into simpler fermentable sugars [16].

The Louisiana Tech University is one among many organizations worldwide engaged in this endeavour, through the work of Dr. James Palmer, in collaborating with fellow professors Dr. Yuri Lvov, Dr. Dale Snow, and Dr. Hisham Hegab [17]. The focus is on non-edible cellulosic biomass, such as wood, grass, stalks, etc, to be converted into ethanol. This approach to produce ethanol can reduce GHG emissions by some 86% over fossils fuels.

The broader field of nanotechnology research into converting biomass into biofuels is growing fast. For example, in 2007 the oil company BP has granted a research fund of $500 million to the University of California, at Berkeley, and the University of Illinois, to explore the conversion of corn, plant material, algae and switch grass into fuel [18].

In the past, Berkeley had used nanotechnology in research for cost-effective solar panels [19]. But, the new Energy Biosciences Institute – EBI created at Berkeley will focus on fuel production with minimum environmental impacts and carbon emissions. A three pronged approach is being employed that begins with technologies for better crop production, improved feedstocks processing and development of new biofuels. The application of this approach aims at developing better feedstocks, breaking down plant material into sugars and their conversion to ethanol. Success along this pathway is expected to lead EBI to .investigate the use of nanotechnology to develop other alternative fuels, such as butanol and renewable hydrocarbon fuels.

Another relevant application of nanotechnology is the use of nano-catalysts for the trans-esterification of fatty esters from vegetable oils or animal fats into biodiesel and glycerol [20]. The nano-catalyst spheres replace the commonly used sodium methoxide. The spheres are loaded with acidic catalysts to react with the free fatty acids and basic catalysts to react with the oils. This approach eliminates several production steps of the conventional process, including acid neutralization, water washes and separations. All those steps dissolve the sodium methoxide catalyst so it can't be used again. In contrast, the catalytic nanospheres can be recovered and recycled. The overall result is a cheaper, simpler and leaner process. In summary, the process claims to be economical, recyclable, to react at mild temperatures and pressures, with both low and high FFA (free fatty acid) feedstock, producing cleaner biodiesel and cleaner glycerol, greatly reducing water consumption and environmental contaminants, and can be used in existing facilities.

8. NANOTECH LIQUID ADDITIVES

All previous presentation and discussion referred to solid nano-particles playing a catalytic role in the obtaining biofuels from algae, landfill methane and biomass. The following segments will examine the practical opportunities that exist for liquid nano-particles or droplets [21]. Consider multifunctional surface active liquid additives, whose lubricity enhancement is achieved via the formation of a monolayer over the surfaces in contact with additized fuel. [22] The treat rate for lubricity is determined by the adsorption saturation concentration. Speculate that the improved detergency and water co-solvency is obtained by the formation of nano emulsions. Also, postulate that the more complete combustion and consequent fuel efficiency increase is the result of the behaviour of nano droplets. These nano droplets result from the surfactant action of the additive in the fuel formulation and the presence of some water in all commercial fuel systems, usually due to evening condensation.

Research by Wulff and colleagues [23] has shown that nano emulsions, which the authors call micro emulsions, with fuel (biofuel included most likely), water and surfactant are:

- Thermodynamically stable and
- Microscopically isotropic, and
- Nano-structured (thus, nano emulsions).

Their research concluded that:

- The use of these nano structures with fuel, water and surfactant is able to break the usual trade off between reduction of soot and NO_x emissions, by achieving them simultaneously, and
- For the same fuel consumption, higher efficiency is obtained.

- Strey and collaborators filed patent applications for what they call micro-emulsions used as fuel [24]. The interpretation offered for the behaviour of stable diesel (and most likely biodiesel)-water-surfactant nano emulsions is as follows:
- The surfactant components -oleic acid and nitrogen containing compounds (amines) - dissolve readily in diesel (and possibly in biodiesel) fuel and bind water to it without stirring;
- The water droplets are as small as a nanometer across, helping stabilize the emulsion
- The result is a "liquid sponge", can be stored indefinitely, like ordinary diesel fuel, without risk of phase separation
- This fuel formulation, when burned, results in the near-complete elimination of soot, and a reduction of up to 80% in nitrogen-oxide emissions
- The surfactant in the formulation also burns without creating emissions beyond water, carbon dioxide and nitrogen

9. PUBLIC CONCERNS OVER NANOTECHNOLOGY: SECURITY, HEALTH AND THE ENVIRONMENT

As with all new technologies, there may be cause to concern about impacts, such as on security, health and the environment. Nanotechnologies have been the subject of many assessments seeking to anticipate possible consequences of their deployment, to humans and to the environment. For instance, the Woodrow Wilson Center carried out a Nanotechnology project [25] from 2005. The project managers said that "manipulating materials at the atomic level can have astronomic repercussions, both positive and negative. The problem is no one really knows exactly what these effects may be." This was the motivation for the Project on Emerging Nanotechnology at the Woodrow Wilson Center.

Another initiative came from the International Risk Governance Council – IRGC's Nanotechnology project [26]. Two expert workshops were held. The first in May 2005 focused on how to frame nanotechnology, its risks and its benefits. A distinction was made between the nanotechnologies of the so-called Frame One (passive or classical technology assessment) and Frame Two (active or the social desirability of innovation). The second, in January 2006, concentrated on identifying gaps in nanotechnology risk governance and developing recommendations for improved risk governance.

A symposium on the subject took place in Zurich in July 2006. A presentation by Ortwin Renn[27] discussed the policy implications of Frame One, referred to in Fig. 4. The fact is that "most people have no clear associations when it comes to nanotech. They expect economic benefits but no revolutionary technological breakthroughs. Risks are often not explicitly mentioned but there is a concern for unforeseen side effects. There is a latent concern about industry, science and politics building a coalition against public interest. And one negative incident could have a major negative impact on public attitudes."

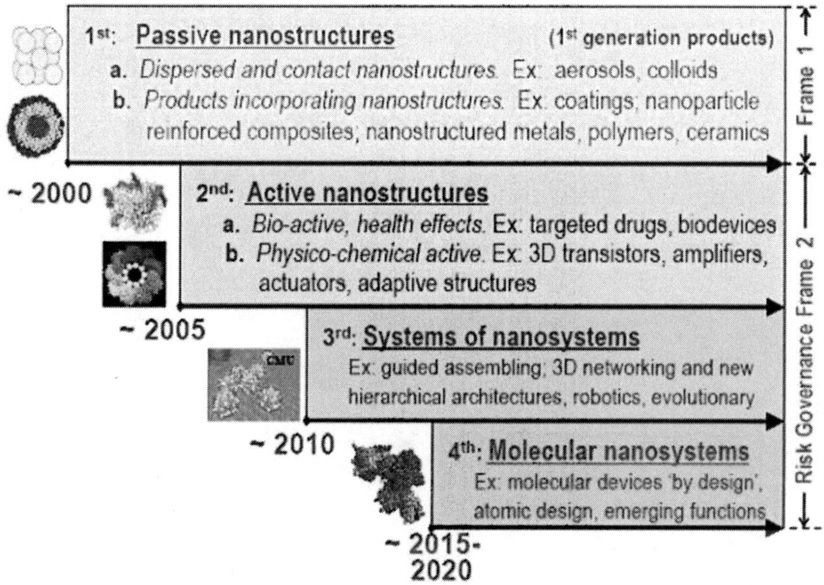

FIGURE 4. Frames of reference of nanotechnology generations

The IRGC's Nanotechnology project concluded [28], among other things, that "communication about nanotechnology's benefits and risks should reflect the distinction between passive and active nano-materials and products, stressing that different approaches to managing risks are required for each. Care should also be taken to ensure that potential societal concerns about the possible impacts of Frame Two active nano-materials do not have the effect of unnecessarily increasing anxiety regarding Frame One products using only passive nanostructures." This is further expounded by Renn [29] as follows: "Frame One passive nanostructures are found, for example, in easy-to-clean surfaces, paints or in cosmetics. Frame Two refers to active nanostructures and molecular systems which could be able to interact actively or could be understood as evolutionary biosystems which change their properties in an autonomous process."

In reality, nanotechnologies are already facing challenges. Man-made nano-materials have been banned by the UK Soil Association from all its certified organic products. The 2008 annual report of the Soil Association of the UK contains the following statement [30]: "The Soil Association published the world's first standards banning nanotechnology. The risks of nanotechnology are still largely unknown, untested and unpredictable. Initial scientific studies show negative effects on living organisms, and three years ago scientists warned the Government that the release of nanoparticles should be 'avoided as far as possible'. There are many parallels with GM in the way nanotechnology is developing, particularly because commercial opportunities have run ahead of scientific understanding and regulatory control. What's more, while nano-substances are being rapidly introduced to the market, there is no official assessment process or labeling of the products – which is even worse than GM.

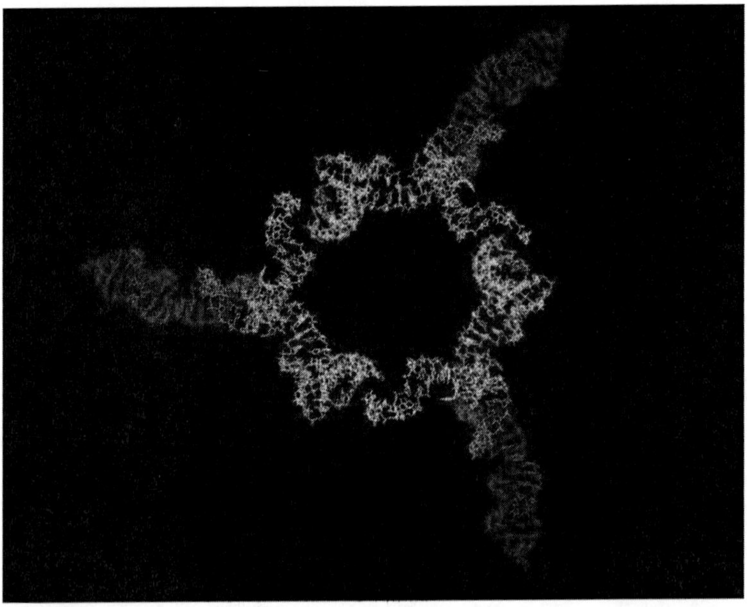

FIGURE 5. RNA nano-particle created by Peixuan Guo, Purdue University [31]

Health and beauty products that use nanoparticles are of concern for their potential toxicity if they get under the skin. Similar concerns exist regarding food and textiles. Definitely, more studies about health and environmental impacts are needed, to alleviate public concerns.

On the other hand, there is so much potential for nanotechnologies to do good, that Frame One and Two assessments should proceed as new applications evolve, including for instance more effective delivery of drugs to fight human and animal disease.

Fig. 5 showing a RNA nano-particle created by Peixuan Guo of Purdue University, illustrates the point. Strands are spliced together from two kinds of RNA – a scaffold and a hunter to find cancer cells. This nano-structure has proven effective against cancer growth in living mice as well as lab-grown human nasopharyngeal carcinoma and breast cancer cells.

10. CONCLUSIONS

Increasing demand for energy services in the decades ahead will require an expanding supply of liquid fuels, despite efforts at improving energy efficiency and diversification of energy systems, including growing use of electricity in transportation. Biofuels have a key role to play in this scenario. However, the future supply of biofuels must be of such a scale that non-food feedstocks and new technologies are intensively employed. Nanotechnologies are primary candidates to play a prominent role in this energy future. They will help bring to markets liquid biofuels, including renewable hydrocarbons, from algae, carbohydrates, fatty esters and biogas. Nanotechnologies will also play a role in augmenting the efficiency of using current and future liquid fuels, especially biofuels, by providing improved combustion of nanodroplets. While there are risks in each and every new technology, the world today is much better equipped to assess risks and act accordingly, that it seems possible to advance nanotechnologies applied to biofuels, without jeopardizing security, public health or the environment. But, the reach of nanotechnologies is vast and goes much beyond biofuels and offer hopes in so many areas, including importantly, human health.

REFERENCES

1. S. C. Trindade, 2010Refining will definitely survive. Pipeline Magazine, 29 August 2010

2. S. C. Trindade, 2010Renewable Energy Perspective- a profitable pathway from oil, Exploration and Processing, Fall 2010 [89Sep.
3. S. C. Trindade, 2010International Biofuels Trade: Issues and Options. International Biofuels Conference_São Paulo, 2628May.
4. G. Santana, S. Quirk, 2009Growing Green: An In-Depth Look at the Emerging Algae Industry, Greener Dawn Research, 22 July, 16p.
5. A. Sustainable, Consensus. Biofuels, 2008Statement from a conference hosted by the Rockefeller Foundation Bellagio Study and Conference Center, Bellagio, Italy, 2428March 2008
6. www.defra.gsi.gov.uk2007In: F.O. Lichts's World Ethanol and Biofuels Report, 4365 and Vol4 17 p.391, Turnbridge Wells, U.K.: F.O. Licht, 2006.
7. da. Carvalho, F. Silva, P. R. da, Brum. T. N. Costa, Santos. dos, 2005Nanotechnology/Nanoscience Knowledge Managament emphasizing nanostructured polymers. Presentation, School of Chemistry, UFRJ, Brazil.
8. S. Borschiver, M. J. O. C. Guimarães, T. N. dos, F. C. Santos, Silva. P. R. C. da, Brum, 2005Patenteamento em Nanotecnologia: Estudo do Setor de Materiais Poliméricos Nanoestruturados., Polímeros: Ciência e Tecnologia, 15n° 4, 245248
9. http://ecolocalizer.com/2009/04/23/nanotechnology-to-aid-the-commercial-viability-of-algal-bio-fuel-production, 2009
10. http://www.qsinano.com/news/releases/2009_02_24.php
11. http://www.ameslab.gov/news/news-releases/nanofarming-technology-extracts-biofuel-oil-without-harming-algae
12. http://www.ameslab.gov/news/news-releases/nanofarming-technology-extracts-biofuel-oil-without-harming-algae
13. http://biomassmagazine.com/articles/2354/dudek-catalyx-nanotech-to-build-landfill-facilities
14. http://journalstar.com/news/local/article_6d5b6a34-e86f-11df-ae58-001cc4c002e0.html2010
15. http://en.wikipedia.org/wiki/Global_warming_potential
16. http://www.sciencedaily.com/releases/2009/10/091008131858.htm
17. Nanotechnology Used In Biofuel Process to Save Money, Environment Science Daily (Oct.2009
18. http://berkeley.edu/news/media/releases/2007/02/01_ebi.shtml
19. U.S. Department of Energy.Berkeley Lab Helios Project. (n.d.) Helios Solar Energy Research Center. Goals and challenges. Retrieved

December 10, 2009from http://www.lbl.gov/LBL-Programs/heliosserc/html/goals.html
20. http://www.public.iastate.edu/~nscentral/news/2007/jun/catilin.shtml
21. Cleaner diesel engines- pouring water on troubled oils, The Economist, June 3rd,201086http://www.economist.com/node/16271415
22. http://www.internationalfuel.com
23. P. Wulff, B. Lada, S. Engelskirchen, R. Strey, 2008Water-biofuel microemulsions. Institute for Physical Chemistry, University of Cologne.
24. http://strey.unikoeln.de/fileadmin/user_upload/Download/WATER__BIOFUEL_MICROEMULSIONS.pdf
25. http://strey.uni-koeln.de/333.html?&L=1
26. R. Strey, et al.2007Microemulsions and use thereof as a fuel. US Patent Application 2007/028507, Feb. 8. http://www.rexresearch.com/strey/strey.htm
27. http://www.loe.org/shows/segments.htm?programID=05-P13-00050&segmentID=3
28. http://www.irgc.org/-Nanotechnology-.html
29. http://www.yasni.ch/ext.php?url=http%3A%2F%2Fwww.irgc.org%2FIMG%2Fpdf%2FOrtwin_Renn_Nanotechnology_Frame_1_Policy_Implications_.pdf&name=Ortwin+Renn&cat=document&showads=1
30. http://www.irgc.org/Policy-Recommendations,188.html
31. http://ec.europa.eu/health/ph_risk/documents/ev_20081002_rep_en.pdf, p.14
32. http://www.soilassociation.org/LinkClick.aspx?fileticket=Moyw3Q7H%2Fp4%3Dtaid=303,22
33. http://www.eng.uc.edu/nanomedicine/Papers/1NCI.pdf

2
CHAPTER

ASSESSMENT OF FEEDSTOCK OPTIONS FOR BIOFUELS PRODUCTION IN GHANA

Francis Kemausuor, Joseph Oppong Akowuah, Emmanuel Ofori

Department of Agricultural Engineering, Kwame Nkrumah University of Science and Technology, Kumasi, Ghana

ABSTRACT

In the wake of climate change and increasing fossil fuel prices, biofuels are becoming attractive to agricultural dependent economies in sub-Saharan Africa and other regions of the world. This study evaluates the energy production potential of biomass resources grown on the available arable agricultural land under two principal scenarios: using 2.5% and 5% of the available arable land for energy crop expansion. Using conservative biofuel yields from crops in the sub-region, a 2.5% of uncultivated arable land dedicated to four traditional crops grown in Ghana namely maize, cassava, sweet sorghum and oil palm could potentially replace 9.3% and 7.2% of transportation fuels by 2020 and 2030 respectively. Using 5% of the uncultivated arable land to cultivate the above four crops and jatropha could potentially produce biofuel to replace 17.3% of transport fuels by 2020 and 13.3% by 2030. In order to enrol such a scheme, government is encouraged to put in place appropriate structures to ensure that, the industry meet international sustainability standards.

Keywords: Biofuels; Feedstock; Ghana

1. INTRODUCTION

As energy demand grows, self-sufficiency in energy requirement has become critical to the socio-economic development of most developing countries. The growth in energy demand due to rapid industrialization and socioeconomic growth require huge investments to meet them. For decades now, fossil oil has contributed immensely to meeting these demands. However, increasing concerns about emissions from fossil fuels and the impending peak oil theory [1-4] have stimulated interest for alternative energy sources. An alternative fuel must be technically feasible, economically competitive, environmentally acceptable, and readily available [5]. Biofuel produced from energy crops and other resources has been touted as one of the viable sources of energy for the foreseeable future in dealing with these worldwide socioeconomic and environmental concerns [6,7]. Global production of biofuel has been growing rapidly in recent years, rising from about 18 billion litres in 2000 to 105 billion litres in 2010 [8]. Despite this exponential increase, biofuel still represents a very small share of global energy consumption. Biofuels provided 2.7% of all global fuel for road transportation in 2010—an increase from 2% in 2009. As usual, supply is dominated by bioethanol, which accounted for approximately 82% of total biofuel production in 2010. Ethanol's contribution is largely responsible for the increase in biofuel production. Together, Brazil and the US contributed about 90% of global ethanol production with the European Union contributing about 53% of all biodiesel in 2010. Global demand for biofuel is projected to increase to about 183.8 billion litres by 2015 with ethanol expected to contribute about 80% of the total [9]. Already, several countries have adopted policies to promote liquid biofuel development led by the US, the EU, Brazil, Canada, Australia and Japan [10]. A growing number of developing countries, including a few African countries have also started to introduce similar policies. Notable countries in Africa with biofuel strategies in place are South Africa, Mozambique, Ethiopia, Senegal and Mali.

In Ghana, a draft bioenergy policy of 2010 [11] aims to substitute national petroleum fuels consumption with 10% biofuel by 2020 and 20% by 2030. The policy intends to make use of the country's vast biomass potential and resources in modern applications for the production of transport fuels and electricity. Biomass already dominates the energy consumption pattern in Ghana, accounting for over 63% of total energy consumed [12] but used principally in traditional cookstoves as firewood and charcoal. Liquid biofuel use is not commercialised in the country but there are plans to begin commercial production as soon as the infrastructure is in place. Plans for biodiesel production are at a more advanced stage as opposed to those for ethanol production. The draft bioenergy policy is open to the commercial scale production of biofuel feedstocks to sustain supply but is not targeting any specific energy crop(s) to meet the demand targets. In the past, jatropha was thought to be the best feedstock due to previous impression that it was drought resistant. But current research has shown that jatropha may not be the wonder crop that it was thought to be [13].

With conflicting reports on the prospects or potential of jatropha and recent revelations that it performs poorly in marginal lands [13], there is the need to perform an analysis of possible biofuel feedstock choices in order to recommend the best choices to investors and policy makers. The aim of this study was to perform an analysis of the technical potential of first-generation biofuels from energy crops in Ghana. The study aims to contribute knowledge to the call from Ghana's Energy Commission for research support to select the best biofuel feedstock options for the country. Knowledge about the technical potentials would enable a good estimation of land occupancy to attain biofuel targets proposed by the Energy Commission. This paper is structured into 5 sections. As part of the introduction, a review of the agricultural resources in the country has been presented. The methodology for conducting the analysis follows the introduction after which the results is presented and discussed. Section 4 presents a brief description of the recipes for a successful biofuel industry in Ghana after which Section 5 summarises the conclusions.

Agricultural Resources in Ghana

Ghana is administratively divided into ten regions with a total land area of approximately 238,537 sq. km. There are 212 Districts/Municipal/Metropolitan Areas (MMDAs), which increased from 138 in year 2000[1]. A recent population census in 2010 estimates Ghana's population at 24.6 million [14]. A regional population density map is shown in **Figure 1**.

The coastal and southern parts have higher population density than the northern parts of the country. The Greater Accra region which also houses the nation's capital has the highest population density whereas the Northern region has the lowest. The Northern region has an estimated 35 persons per square kilometre with vast uncultivated arable lands. In relation to agriculture, there are six agro-ecological zones in Ghana defined on the basis of climate, reflected by the natural vegetation and influenced by the soil type (**Figure 2**).

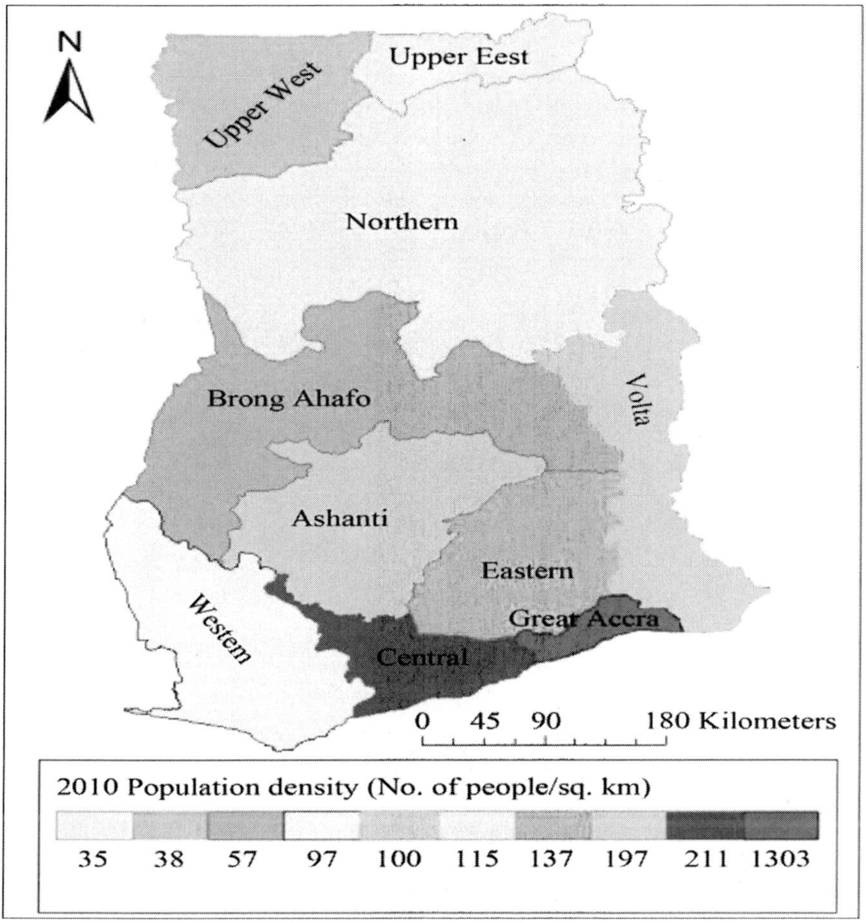

FIGURE 1. Regional population density map of Ghana for year 2010 [14].

These are Rain Forest, Deciduous Forest, Transitional Zone, Coastal Savannah, Guinea Savannah and Sudan Savannah. The characteristics of the various agro-ecological zones are shown in **Table 1**. Annual rainfall ranges from about 800 mm along the coastal savannah to 2200 mm in the Rain forest. Annual average temperatures range from 26.1°C near the coast to 28.9°C in the northern parts of the country. About 155,000 sq. km., approximately 65% of the total land area, is classified as agricultural land area [15]. The agricultural sector is characterized by a large number of dispersed small-scale producers, employing manual cultivation techniques dependent on rainfall with little or no purchased inputs but providing over

90% of the food needs of the country [16]. Farming systems vary with the six agro-ecological zones (**Table 1**). However, certain general features are discernible throughout the country. According to [17], crop production in Ghana is hampered by land degradation, improper field development, use of low-yielding varieties, lack of organised seed production and distribution systems, and inadequate storage structures. Situated in a tropical climate, Ghana has the advantage of growing the most desirable energy crops for biofuel production. Table 2 shows a summary of key characteristics of some of these crops.

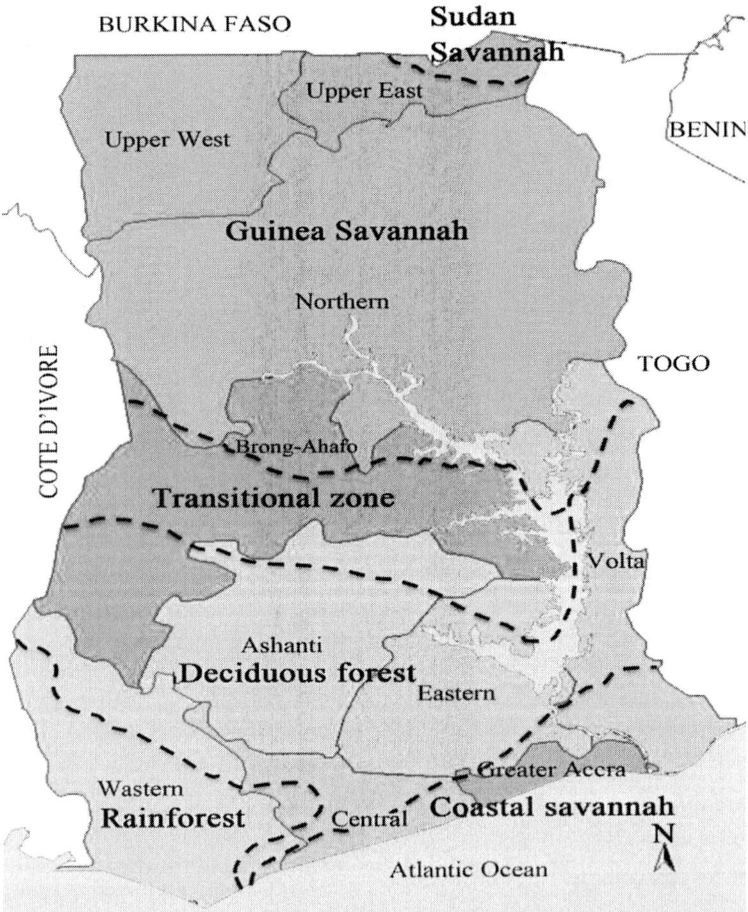

FIGURE 2. Agro-ecological zone classification in Ghana.

TABLE 1. Characteristics of agro-ecological zones in Ghana.

Zone	Area (1000 ha)	Rainfall (mm/yr)	Length of growing season (days)	Dominant land use systems	Main food crops
Rain forest	750	2200	Major season: 150 - 160 Minor season: 100	Forest, plantations	Cassava, plantain, maize, oil palm
Deciduous forest	740	1500	Major season: 150 - 160 Minor season: 90	Forest, plantations	Cocoa, cassava, plantain, maize, oil palm
Transition Zone	6630	1300	180 - 200	Annual food and cash crops	Maize, cassava, yam, taro (cocoyam), plantain, groundnut, cowpea, maize
Guinea savannah	14,790	1100	180 - 200	Annual food and cash crops, livestock	Sorghum, maize, groundnut, millet, yam, cowpea, maize
Sudan Savannah	190	1000	150 - 160	Annual food crops, livestock	Millet, sorghum, cowpea, groundnut, yam, maize
Coastal savannah	580	800	Major season: 100 - 110 Minor season: 50	Annual food crops	Cassava, maize

Source: [18].

Table 2. Overview of principal biofuel feedstocks.

Feedstock	Conditions for best performance	Fuel yield per hectare of crop (l/ha)
Sugarcane	Up to 1600 m above sea level	5800 l/ha in Brazil, 4000 l/ha for Africa
Corn (maize)	Grows everywhere but often needs irrigation	3000 l/ha in advanced agricultural systems, 700 l/ha for Africa due to poor yields
Sweet sorghum	Dryer tropics and temperate regions up to 2500 m altitude	3000 - 6000 l/ha
Cassava	Tropical climates up to 1000 m altitude	5400 l/ha advanced agricultural systems, 1750 l/ha in Africa due to low efficiency in current production methods
Oil Palm	Humid tropic areas up to 700 m altitude	6000 l/ha in Malaysia, Africa plantations average less than half
Jatropha	Tropical & semi-arid regions at altitudes up to 500 m. Rainfall from as low as 300 mm	400 - 2200 l/ha of pure plant oil

Source: Compiled from [19].

With regards to Food security, 5% of Ghana's population are food insecure. The food security problem exists in all the agro-ecological zones and in both urban and rural areas. There have been many agricultural projects implemented to address the food security problem particularly in relation to stable food crops such as maize, rice, sorghum and soybean. Root and tuber crops such as cassava, yam and cocoyam have also been considered. The government is introducing block farm programmes to make it possible for government sponsored mechanisation and extension services to cover large number of farmers. The Ministry of Food and Agriculture has also set up the National Food Buffer Stock Company (NAFCO) to ensure food security and to insulate farmers against losses resulting from anticipated increases in production.

2. METHODOLOGY

2.1. Crop Cultivation Pattern in Ghana

In 2010, approximately 7.8 million hectares, or 57.6% of the available arable land for agriculture in Ghana was under agricultural cultivation [16] leaving approximately 5.8 million (or 42.4%) which was classified as uncultivated. **Table 3** shows the cultivated area and production of major agricultural crops in Ghana in the year 2010. Potential crops for bioenergy that dominate the land area are maize, cassava, oil palm and sorghum. Other potential crops such as soybean, sunflower, sugarcane and coconut occupy less land areas, Maize is the most cultivated non-export crop in Ghana followed by cassava. In 2010, approximately 1.9 million tonnes of maize was harvested from an area of 992 thousand ha. Maize is produced in all districts in Ghana, making it the most popular food crop in the country. Over 13.5 million tonnes of cassava was produced from an area of 875 thousand hectares in the same year. About 2 million tonnes of oil palm fruit and 0.32 million tonnes of sorghum were produced in 2010 from an area of 360 thousand hectares and 253 thousand hectares respectively. With regards to the trend in the cultivated area of these four crops, the pattern shows a fairly even trend over the last decade, with marginal increase or decrease in some cases (**Figure 3**). There were notable increases in the cultivated areas of oil palm (96%) and maize (39%) and a fairly even trend in the cultivated area for sorghum (**Figure 3**).

TABLE 3. Cultivated area and production of major crops in Ghana, 2010.

Crop type	Area harvested 2010 (ha)	Production 2010 (t)	Yield (t/ha)
Cocoa beans	1,625,000	632,037	0.39
Maize	991,669	1,871,700	1.89
Cassava	875,013	13,504,100	15.44
Yams	384,942	5,960,490	15.48
Groundnuts	353,376	530,887	1.50
Oil palm fruit	352,800	2,004,300	5.68
Plantains	327,870	3,537,730	10.79
Sorghum	252,555	324,422	1.28
Cocoyam	205,342	1,354,800	6.60
Rice, paddy	181,228	491,603	2.71
Millet	176,600	218,952	1.24
Vegetables	165,010	818,615	4.9610

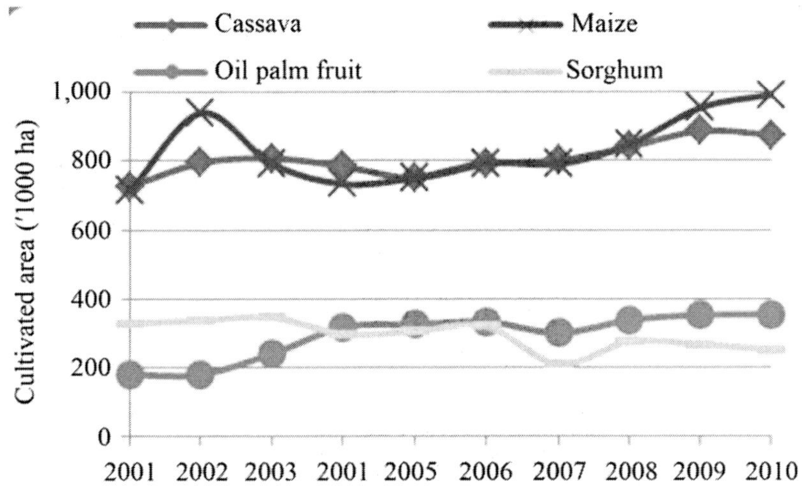

FIGURE 3. Pattern of land use for selected crops in Ghana. Source: Data from [20].

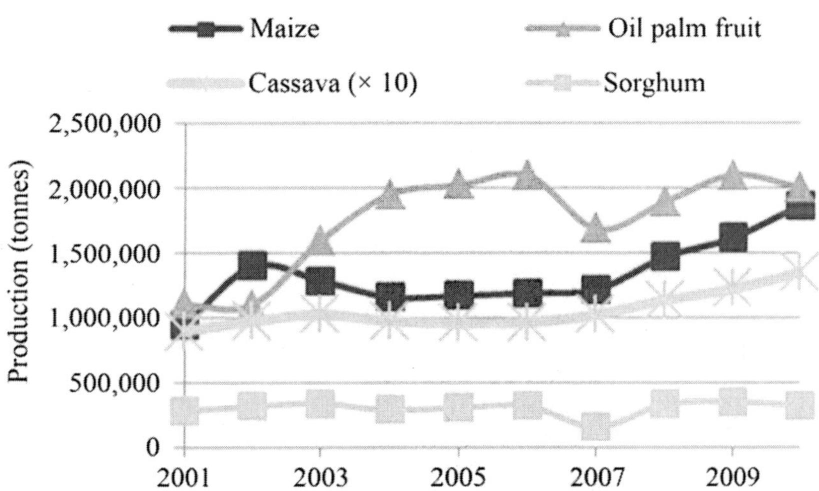

FIGURE 4. Pattern of production of selected crops in Ghana. Source: Data from [20].

Cassava is the dominant food crop in Ghana in terms of production output and has remained so for the last decade (**Figure 4**). It is cultivated in eight of the ten regions in the country. The large production of cassava is partly the result of a cassava improvement programme— the "Root and Tuber Improvement and Marketing Programme" (RTIMP)[2]—supported by the International Fund for Agricultural Development (IFAD) and the Government of Ghana (GoG) through the Ministry of Food and Agriculture. Generally, oil palm cultivation is carried out in varying scales such as smallholder farms, and medium to large-scale plantations. Of the 360 thousand hectares of oil palm cultivated in 2010, 60 thousand hectares were commercial plantations and the rest (over 80%) are small-scale farms scattered over the southern parts of the country [16].

Based on land area already occupied by potential bioenergy crops, four crops namely, maize, cassava, oil palm, and sorghum have been considered in this paper as possible crops for biofuel production in Ghana. Production trends were similar to the area cultivated for these crops. There are notable increases in the production of maize (99.5%), oil palm (81%) and cassava (51%) which corresponds somewhat to the increases in their cultivated area, indicating that there has been little improvement in the yields of these four crops over the period.

2.2. Scenarios Development

Two principal scenarios were adopted in the assessment of feedstock options for biofuel production in Ghana. A percentage of the uncultivated arable agricultural land available in Ghana is dedicated to growing the selected potential bioenergy crops in the different scenarios to determine the resulting biofuel yields. The biofuel produced in the different scenarios is compared to projected petroleum fuel demand for transportation to determine the percentage of fuel it can replace. Crop data used for the assessment was crop production data from the Statistics, Research and Information Directorate (SRID) of Ministry of Food and Agriculture and also from the Food and Agriculture Organisation. Fuel consumption data and projections were from the National Petroleum Authority and the Energy Commission of Ghana. In the analysis, a portion of uncultivated arable agricultural land in Ghana is dedicated to cultivating the four potential energy crops identified, i.e. maize, cassava, oil palm and sweet sorghum in a proportion equal to the proportion in which they were cultivated in 2010. Besides being the most cultivated bioenergy crops in terms of land area, Ghanaian farmers are familiar with their cultivation and the crops are also well suited to the climatic and soil conditions in Ghana. With regards to sorghum, we expect the "sweet sorghum" variety to be cultivated for energy crop purposes. In the first scenario, 2.5% of unused agricultural land (or 144,541 ha) is used to expand the cultivation of these four crops for biofuel purposes. In 2010, approximately 2480 hectares of land was dedicated to these four crops in Ghana. Of this amount, maize and cassava occupied about 75% as shown in **Table 4**. In scenario one, the assumption was made that, 2.5% of uncultivated agricultural land in 2010, which is 144,541 ha is shared among the four crops in the percentages shown in **Table 4** which is the percentage occupied by each of these crops as a total of the area they occupied in 2010. Cultivation is assumed to start gradually and the 2.5% or 144,541 hectares is expected to be fully cultivated in the year 2020. Beyond 2020, no additional land is expected to be allocated to biofuel crops but rather expect further research conducted to improve upon the conservative yields that were used in the initial analysis up to year 2030. The timelines adopted is in line with the Energy Commission's adoption of these dates to realize biofuel blends in transportation fuels in Ghana.

Having allocated land space to each of these crops, biofuel production potential was estimated based on conservative yields achieved in most African countries as reported by [19] and also presented in **Table 2**. These yield figures are based on current agricultural practices in African countries with low irrigation and low fertiliser input. For example, whereas ethanol from maize could yield as much as 3000 litres/hectare in developed agricultural systems, the reported figure for Africa is 700 litres/hectare due to low yields. Also in the case of cassava, about 5400 litres/hectare could be realised in countries such as Thailand but only 1750 litres/hectare is reported for Africa. In order to make a case for research into agronomic improvements, the amount of land space that could be saved is estimated if yields in Ghana were to improve comparable to the current yield in advanced agricultural economies and assuming the same amount of biofuel is to be produced. The resulting biofuel yields are then compared to transportation fuel demand projected for the years 2020 and 2030.

Table 4. Agricultural land allocation to crops in Scenario 1.

Crop	Land area in 2010 (ha)	Fraction of total land area occupied by 4 crops	Portion allocated from 2.5% uncultivated land
Maize	992	0.4	57816.3
Cassava	875	0.35	50589.2
Oil palm	360	0.15	21681.1
Sorghum	253	0.1	14454.1
TOTAL	2480	1	144540.7

The second scenario considered 5% of the uncultivated agricultural land available in Ghana dedicated to growing the identified potential bioenergy crops for biofuel. Scenario 2 differs from Scenario 1 with the addition of jatropha to the four original crops. In the land allocations, 2.5% of the uncultivated land is still allocated to the four crops in the same proportion as in Scenario 1 and an additional 2.5% dedicated to cultivating only jatropha. The 2.5% allocation to jatropha is admittedly high; but this is based on the fact that cassava and especially oil palm are, to a large extent, not cultivated in the northern parts of the country where there is very low population density with large uncultivated lands. The northern parts of the country occupy close to 40% of Ghana's land size. Annual rainfall amounts of 800 to 1100 mm make it feasible to cultivate jatropha in the northern parts of the country. It was therefore assumed that, the bulk of the land dedicated to jatropha in Scenario 2 would be sourced from the northern parts of the country where jatropha is already cultivated on commercial scale.

Again the assumption was made that biofuels produced in both scenarios are to be used locally while the trend in food crop production remains unchanged. It is expected that, improved agronomic practices in the energy crop plantations would trickle down to smallholder food crop programmes which should bring about marginal increases in yields.

3. RESULTS AND DISCUSSION

This section presents the results and discussion of the assessment of feedstock options for biofuel production in Ghana.

3.1. Results from Scenario 1

Table 5 presents the results for biofuel production from Scenario 1. Using conservative yields for biofuels from the various crops under consideration, about 303 million litres can be produced from 2.5% of uncultivated agricultural land in Ghana. This is made up of about 219 million litres bioethanol and 84 million litres biodiesel. About 85% of the total land allocated is dedicated to bioethanol crops and only 15% to biodiesel crops. **Table 6** shows the amount of land area that can be saved if the same amount of biofuel is produced but with high crop yields. In **Table 6**, it is assumed that the country intends to produce the same amount of biofuel (303,194,517 litres) as in **Table 5** but invests in agronomic research that improves yields comparable to the current levels in advanced agricultural economies. It also further takes into consideration, the fact that, high agricultural productivity is achieved with inputs that enhance crop growth and with irrigation schemes where necessary that ensure that crops are not rainfall dependent. Using yield figures from **Table 2**, the targeted biofuel is produced with 57,756 hectares of land and saves over 86,000 hectares of land space in the process compared with conservative yields.

TABLE 5. Biofuel production under Scenario 1.

Crop	Land allocated (ha)	Biofuel yield (l/ha)	Biofuel production (l)	Fuel type
Maize	57816.3	700	40,471,410	Bioethanol
Cassava	50997.2	3000	152,931,600	Bioethanol
Sweet Sorghum	14745.5	1750	25,804,625	Bioethanol
Oil Palm	20981.7	4000	83,926,800	Biodiesel
Total	**144540.7**		**303,134,435**	

TABLE 6. Land area saved with high yields from Scenario 1.

Crop	Desired biofuel production (l)	High biofuel yield (l/ha)	New land area to cultivate (ha)	Land area saved
Maize	40,471,396	3000	13,490	44,326
Cassava	152,991,668	6000	25,499	25,499
Sweet Sorghum	25,804,595	5400	4779	9967
Oil Palm	83,926,858	6000	13,988	6994
Total	**303,194,517**		**57,756**	**86,786**

3.2. Results from Scenario 2

Table 7 shows biofuel production in Scenario 2 where an extra 2.5% of the land is used for jatropha cultivation alone. In this scenario, about 563 million litres of biofuel is produced in the country. Bioethanol production under this scenario remains at the same level as in Scenario 1, at 219 million litres. However, biodiesel production increases from the original 84 million litres in scenario 1 to 344 million litres. Again in Scenario 2, if the same amount of biofuel mentioned above is to be produced from high yielding crops, the land savings expected is estimated. In this case achieving high yields from biofuel crops would save the country about 113 thousand hectares of land (**Table 8**) which is more than a third of the land used in Scenario 2 for conservative yields. To achieve such high biofuel yields, cassava, oil palm, sweet sorghum and maize yields would have to increase two fold, three-fold, four-fold and five-fold respectively compared to existing yields.

TABLE 7. Biofuels production under Scenario 2.

Crop	Land allocated (ha)	Biofuel yield (l/ha)	Biofuel production (l)	Fuel type
Maize	57,816	700	40,471,396	Bioethanol
Cassava	50,997	3000	152,991,668	Bioethanol
Sweet Sorghum	14,745	1750	25,804,595	Bioethanol
Oil Palm	20,982	4000	83,926,858	Biodiesel
Jatropha	144,541	1800	260,173,260	Biodiesel
Total	**289,081**		**563,367,777**	

TABLE 8. Land area saved with high yields in Scenario 2.

Crop	Desired biofuel production (l)	High biofuel yield (l/ha)	New land area to cultivate (ha)	Land area saved
Maize	40,471,396	3000	13,490	44,326
Cassava	152,991,668	6000	25,499	25,499
Sweet Sorghum	25,804,595	5400	4779	9967
Oil Palm	83,926,858	6000	13,988	6994
Jatropha	260,173,260	2200	118,261	26,280
Total	**563,367,777**		**176,016**	**113,066**

3.3. Fossil Fuel Replacement

Petrol and diesel consumption for the years 2020 and 2030 was estimated using correlation with population growth to determine portions of road transport fuels that can be substituted with biofuel. Demand for petrol and diesel is expected to reach 1.28 billion litres and 1.97 billion litres respectively by the year 2020. This increases to 1.66 billion litres and 2.56 billion litres for petrol and diesel respectively in 2030. **Table 9** shows the percentage of petroleum fuels that can be substituted with biofuel produced in Scenarios 1 and 2. In 2020, there is a potential of replacing 9.3% of overall diesel and petrol demand with biofuel from Scenario 1. In Scenario 2, there is the potential to substitute 17.3% of petroleum fuels. In Scenario 2, Ghana can substitute 17.1% of petrol with ethanol and 17.5% of fossil diesel with biodiesel. In 2030, Ghana would be able to substitute 7.2% of its petrol and diesel consumption in Scenario 1 and 13.3% in Scenario 2.

3.4. Discussion

Ghana's draft bioenergy policy of 2010 [11] aims for the substitution of national petroleum fuels consumption with biofuel by 10% by 2020 and 20% by 2030. If Ghana were to adopt Scenario 1, the country would be able to replace 9.3% and 7.2% respectively by 2020 and 2030.

TABLE 9. Transport fuel substitution with biofuel from conservative yields.

	Petroleum consumption (l)	Biofuel production (l)	Petroleum replaceable (%)
Scenario 1 (2020)			
Bioethanol/Petrol	1,280,303,721	219,207,635	17.1
Biodiesel/Diesel	1,970,649,110	83,926,800	4.3
Total	**3,250,952,831**	**303,134,435**	**9.3**
Scenario 2 (2020)			
Bioethanol/Petrol	1,280,303,721	219,207,635	17.1
Biodiesel/Diesel	1,970,649,110	344,100,060	17.5
Total	**3,250,952,831**	**563,307,695**	**17.3**
Scenario 1 (2030)			
Bioethanol/Petrol	1,666,282,480	219,207,635	13.2
Biodiesel/Diesel	2,564,749,310	83,926,800	3.3
Total	**4,231,031,790**	**303,134,435**	**7.2**
Scenario 2 (2030)			
Bioethanol/Petrol	1,666,282,480	219,207,635	13.2
Biodiesel/Diesel	2,564,749,310	344,100,060	13.4
Total	**4,231,031,790**	**563,307,695**	**13.3**

In effect, Scenario 1, which uses 2.5% of uncultivated arable land at conservative yield estimates, comes close to meeting the 2020 target but fall short of the 2030 target. A 2.5% of uncultivated land space can only meet EC targets if there is a significant increase in crop yields. For both target years, i.e. 2020 and 2030, Scenario 2 presents relatively high percentage replacements. However, in terms of actual energy content, more bioethanol and biodiesel would be required to meet its equivalent petroleum fuel product. This is because the energy content of bioethanol is about two thirds that of petrol and biodiesel has an energy content of about 12% less than petroleumbased diesel fuel on a mass basis [21]. Scenario 2 serve a better option in both target years for meeting the targets of the draft bioenergy policy.

The decision to increase cultivation of certain crops would demand better technological input into the cultivation of crops and in the end, enhance food crop production in the country. Already, existing bioenergy companies in the country are investing into heavy agricultural machinery and agronomic improvements in order to boost their yields and improve on the economics of production. It is therefore anticipated that, the spill over effects would enhance crop yields from small scale farmers as they learn to embrace new and improved variety of crops. Increased agricultural activities, even with high mechanization would create more employment for the country and the extra biofuel could be exported to other countries in the early stages until the infrastructure is ready to support biofuel consumption in Ghana. Presently, the infrastructure for retailing biofuel blends are not in place as biofuel consumption is not mandated by government.

In order to achieve targets of land use, energy and land use regulators, i.e. the Energy Commission and Lands Commission, must ensure that regulatory systems are enforced in order to protect the environment and livelihood. A weakened system and/or regulator would expose the system to exploitation and the perceived benefits would not be achieved. Land acquisition guidelines which factors in all stakeholders (and not chiefs alone as the case is presently), should be in place to ensure a wider participation. Representatives of community members as well as district agricultural and planning officers should be actively involved. Once the structures are in place and lands are demarcated, the Lands Commission must ensure that only lands meant for energy crop production are used for the purpose.

The local production and use of biofuel as substitute for fossil-based fuels offers many attractive benefits for Ghana, but they could also impact negatively if not managed properly. Even though a detailed study on the impact of biofuels falls beyond the scope of this paper, there are notable points that must be taken into consideration in any serious biofuel programme of the kind suggested in this paper. The production of biofuel in Ghana presents an opportunity for local and foreign investments in Ghana, as well as offers the opportunity to reduce crude oil import and reduce the country's vulnerability to supply disruptions. Pursuing Ghana's biofuel potentials would boost agricultural and technological development and create jobs that would enhance the quality of life. The cultivation, harvesting and processing of energy crops could create an additional income avenue for farmers especially during the off-season when they are not engaged in any farming activities. Besides the financial savings that biofuel production of this scale may offer the country, there could also be environmental benefits which include the offset of greenhouse gases (GHGs) associated with the burning of fossil fuels. Exactly how much GHG savings would be realised is still debatable as there is no global agreement on the issue as different computing methodologies have yielded different results.

Notwithstanding the benefits, such a large scale production of energy crops for biofuel could bring with it negative socio-economic and environmental implications. First of all, wages and working conditions could be poor [22,23] and jobs may be given to people living outside plantation communities which could create conflicts. There is also the fear of environmental degradation if regulations are not enforced. This calls for effective regulatory and monitoring agencies actively engaged with all stakeholders.

4. RECIPE FOR SUCCESS

In order to realise a successful biofuel industry, there is the need for important policy issues to be put in place and for government to support the industry, especially during its nascent stage. Some of the more critical issues include:

4.1. Enactment of Policies and Regulatory Frameworks

Government support, legislation and regulation are critical to a successful biofuel industry in Ghana. Even though Ghana has targets for overall biofuels consumption in the near future, there are no mandates in place for biofuel blends (e.g. E5, E10, B5, B10, etc.) which could discourage investment plans, especially by the private sector. It could also lead to a situation where companies might be forced to export seeds/oil with a justification that there is no mandate for the use of biofuel. In order to succeed in introducing biofuel consumption in larger quantities in Ghana, the Energy Commission must set blending ratios similar to what countries such as Brazil has done. For a programme of this nature, biofuel agencies or parallel organisations should be established to oversee biofuel development programmes. Such an agency would be able to determine the appropriate support that the industry needs through interactions with various stakeholders and identify the needed infrastructure necessary to attract local and foreign investors into the industry. Clear responsibilities for Environmental and Social Impact Assessments (ESIAs) should be given to the Environmental Protection Agency. ESIAs will be crucial for ensuring socio-economic and environmental sustainability of biofuel projects. Strong government regulatory agencies will be required to undertake ESIAs and also to force their implementation. One critical issue here is the mapping of lands for food and energy crops, as well as for other purposes. Land use maps must be prepared to cover all agro-ecological zones in the country using GIS as appropriate, in order to lessen or eliminate exploitation of food crop lands by biofuel producers. Such maps should inform decision makers on which areas in the country must be reserved only for food crop production and which areas could be used for energy crop cultivation. This would go a long way towards ensuring that poor farmers are not deprived of their farm lands in the biofuel boom.

4.2. Support for R & D and Technology Transfer

The process of transferring biofuel technology and expertise involves the flow of knowledge, experience and equipment among different stakeholders, including governments, the private sector, financial institutions, NGOs, research and educational institutions and labour unions

[23]. One of the biggest challenges, for biofuels in Ghana, at the moment has to do with achieving high yields from energy crops and this can be done through partnership with the countries that have achieved higher yields in biofuel feedstocks, such as Brazil for sugarcane, Malaysia and Indonesia for oil palm and Thailand for cassava. These countries have already gone through the experimental stages of agronomical development and there are opportunities for Ghana to partner with them to transfer the knowledge acquired to the country. Apart from improvements in feedstock, processing equipment is another important area that needs development. Biofuel plant providers could build plants on a "build-operatetransfer" basis in partnership with Ghana in the beginning. This would bring about improved local capacity to address the technology issue in future projects.

For biofuel technology transfer to be successful, government must be actively involved. Government support for technology transfer may include direct financial incentives, including grants for Research, Development and Demonstration (R, D & D). Policies supportive of international joint ventures would also help provide domestic companies in Ghana with access to intellectual property owned by international companies. With a natural favourable climate for energy crop production, partners in such joint ventures might contribute host sites for demonstrations and first commercial plants, as well as avenues for entering local biofuel markets. R & D support from the research community is needed in developing high yielding energy crops, standards for blending fuels, the development and improvements in production technology and processing, market development and consumer use. It is important that R & D is linked to technology transfer in such a way that it feeds directly into the scale up of biofuel in the country. R & D efforts need to tackle the practical problems encountered by biofuel technology transfer, not just at the technological level, but at all the critical levels as dictated by the experience from the field. An effective R & D programme should really cover Research, Development, Demonstration and Diffusion (R, D, D & D) and the activities should be closely linked with innovation in the market and education both in its formal and informal/non-formal aspects.

4.3. Ensuring Social Equity and Ecological Sustainability

The production of biofuel in Ghana must be done to ensure social equity. Incentives should be provided for the establishment of small scale biofuel conversion technologies in rural communities so that higher-value products may be produced and sold directly from the rural industries. Thus, Prescribed Obligatory Purchases (POPs) requiring private biofuel companies to purchase stipulated percentages of their feedstock or crude oil requirements from local farmers should be enshrined in the national biofuel policy. Another mechanism that could be used to address multiple objectives for the direct benefit of producers at the lowest and local levels is the Biofuel Export Levy (BEL). The objectives of the BEL could include investment in social services, particularly educational and health facilities, in biofuel producing areas; funding for R & D in improving yields of indigenous feedstocks and purchasing of improved feedstock varieties for small-scale local producers. A BEL could be set at a fixed percentage, up to about 10%, of revenues from biofuel exports and this should be explicitly shared among the different objectives agreed for the BEL.

The production and use of biofuel may have serious environmental impacts such as the use of large amounts of water, destruction of forests, reduction in food production and increase in soil degradation. It is important for Ghana to recognise the ecological challenges associated with biofuel and to address these challenges in order to ensure a sustainable industry. In this regard, it is necessary to put in place sustainability standards to address specific environmental and social issues such as soil erosion; damage to water and soil from the application of pesticides and fertilisers; increased use of freshwater sources; ecological impacts of monoculture crop plantations; the loss of biodiversity and wildlife habitat; potential impacts on agricultural and rural incomes; access to biofuel markets by small landholders and indigenous groups; job availability and quality; potential use of child labour; and access to education and healthcare for workers. The residues generated during the harvesting and processing of the crops could be used to produce biogas which can be used as cooking fuels, thereby reducing deforestation and improving public health.

5. CONCLUSION

Ghana has the potential to cultivate energy crops for biofuel production. In this assessment, two scenarios were used: Scenario 1 had 2.5% of uncultivated arable land dedicated to only four potential energy crops and Scenario 2 builds on Scenario 1 with an additional 2.5% of uncultivated arable land dedicated to only jatropha. The 5% of the uncultivated agricultural land was allocated to the energy crops in the following percentages: 21.2% for maize, 2.5% for sweet sorghum, 18.7% for cassava, 7.6% for oil palm and 50% for jatropha. Using conservative biofuel yields of 700 l/ha for maize, 3000 l/ha for cassava, 1750 l/ha for sweet sorghum, 4000 l/ha for oil palm and 1800 l/ha for jatropha, a 2.5% of uncultivated arable land dedicated to the four crops presented in Scenario 1 would yield 303 million litres of biofuel. Using the same yields in Scenario 1 and adding 2.5% of uncultivated arable land for jatropha cultivation yields 563 million litres of biofuel. The volume of biofuel produced in Scenario 1 could substitute 9.3% of transportation fuels by 2020 and 7.2% by 2030. In the second scenario, Ghana could replace 17.3% of transportation fuels by 2020 and 13.3% by 2030. At such conservative yield, the country would have to dedicate more land to energy crops in order to meet projections for biofuel demand. Other ways of meeting demand are to raise yields of these crops and/or supplement with production from lignocellulosic materials.

REFERENCES

1. G. Maggio and G. G. Cacciola, "When Will Oil, Natural Gas, and Coal Peak?" Fuel, Vol. 98, 2012, pp. 111-123.
2. M. Aftabuzzaman and E. Mazloumi, "Achieving Sustainable Urban Transport Mobility in Post Peak Oil Era," Transport Policy, Vol. 18, No. 5, 2011, pp. 695-702.

3. A. Zecca and L. Chiari, "Fossil-Fuel Constraints on Global Warming," Energy Policy, Vol. 38, No. 1, 2010, pp. 1-3.
4. N. A. Owen, O. R. Inderwildi and D. A. King, "The Status of Conventional World Oil Reserves-Hype or Cause for Concern?" Energy Policy, Vol. 38, No. 8, 2010, pp. 4743-4749.
5. J. Singh and S. Gu, "Biomass Conversion to Energy in India—A Critique," Renewable and Sustainable Energy Reviews, Vol. 14, 2010, pp. 1367-1378.
6. J. J. Cheng and G. R. Timilsina, "Status and Barriers of Advanced Biofuel Technologies: A Review," Renewable Energy, Vol. 36, No. 12, 2011, pp. 3541-3549.
7. A. Demirbas, "Biofuels Securing the Planet's Future Energy Needs," Energy Conversion and Management, Vol. 50, 2009, pp. 2239-2249.
8. Worldwatch Institute, "Biofuels Make a Comeback Despite Tough Economy," Worldwatch Institute, 2011. http://www.worldwatch.org/biofuels-make-comeback-despite-tough-economy
9. HART/GBC, "Global Biofuels Outlook 2009-2015," Hart Energy Consulting/Hart's Global Biofuels Center, Texas, 2009.
10. C. Mandil and A. Shihab-Eldin, "Assessment of Biofuels Potential and Limitations," A Report Commissioned by the International Energy Forum, 2010. http://www.ief.org/_resources/files/content/news/presentations/ief-report-biofuels-potentials-and-limitations-february-2010.pdf
11. Energy Commission, "Draft Bioenergy Policy of Ghana," Energy Commission, Accra, 2010.
12. F. Kemausuor, G. Y. Obeng, A. Brew-Hammond and A. Duker, "A Review of Trends, Policies and Plans for Increasing Energy Access in Ghana," Renewable and Sustainable Energy Reviews, Vol. 15, No. 9, 2011, pp. 5143- 5154.
13. Africa Biofuel Network, "Biofuels—A Failure for Africa," 2010. http://www.africanbiodiversity.org/sites/default/files/PDFs/Biofuels%20-% 20A% 20Failure%20for% 20Africa%20(ABN,%20Dec% 202010).pdf
14. Ghana Statistical Service, "Population and Housing Census Summary of Final Results, 2012," 2012. http://www.statsghana.gov. gh/docfiles/2010phc/2010_popution_and_housing_census(view_summary_of_final_results).pdf

15. Food and Agricultural Organisation, "Ghana Overview, 2009." http://faostat3.fao.org/home/index.html#VISUALIZE_BY_AREA
16. Ministry of Food and Agriculture, "Agriculture in Ghana: Facts and Figures 2010," Ministry of Food and Agriculture, Statistics, Research and Information Directorate (SRID), 2011.
17. M. H. Duku, S. Gu and E. B. Hagan, "A Comprehensive Review of Biomass Resources and Biofuels Potential in Ghana," Renewable and Sustainable Energy Reviews, Vol. 15, No. 1, 2011, pp. 404-415.
18. Food and Agricultural Organisation, "Ghana Aquastat 2005." http://www.fao.org/fileadmin/user_upload/aquastat/pdf_files/GHA_tables.pdf
19. S. Sielhorst, J. W. Molenaar and D. Offermans, "Biofuels in Africa: An Assessment of Risks and Benefits for African Wetlands," Wetlands International, Wageningen, 2008.
20. Food and Agricultural Organisation, "Crop production Ghana 2010." http://faostat.fao.org/site/567/default.aspx
21. K. M. Rahman, M. Mashud, M. Roknuzzaman and A. Al Galib, "Biodiesel from Jatropha Oil as an Alternative Fuel for Diesel Engine," International Journal of Mechanical & Mechatronics, Vol. 10, No. 3, 2010, pp. 1-6.
22. A. Milbrandt, "Assessment of Biomass. Resources in Liberia," Prepared for the US Agency for International Development (USAID) under the Liberia Energy Assistance Program (LEAP), Technical Report NREL/TP-6A2- 44808, 2009.
23. Worldwatch Institute, "Biofuels for Transport: Global Potential and Implications for Energy and Agriculture," Earthscan Publishing, 2007.

3
CHAPTER

KINETIC STUDY ON PALM OIL WASTE DECOMPOSITION

Khan Zakir[1], Yusup Suzana[1], M. Ahmad Murni[1], Uemura Yoshimitsu[1], S. Chok Vuoi[2], Rashid Umer[1] and Inayat Abrar[1]

[1] *Universiti Teknologi PETRONAS, Perak, Malaysia*

[2] *Platinum Energy Sdn. Bhd., Kuala Lumpur, Malaysia*

1. INTRODUCTION

Malaysia is the largest producer of palm oil and contributes 43% of worldwide production (Shuit et al., 2009). Beside palm oil, palm oil industry generated 169.72 million metric tons solid wastes which contribute 85.5% of total biomass waste produced in the country (Khan et al., 2010). This huge amount of wastes can be converted into valuable chemical feed stocks and fuels due to environmental problems associated with conventional fossil fuels.

It is well known that lignocellulosic biomass mainly consists of hemicellulose, cellulose and lignin. The usual proportions (wt%) vary as 40-50% cellulose, 20-60% hemicellulose and 10-25% lignin (Yang et al., 2007). The thermal decomposition of these individuals is important since they influence the basics of thermochemical conversion processes such as pyrolysis, combustion and gasification. Decomposition of these components is intensively studied in the literature. Demirbas et al. (2001) observed the ease of lignocellulosic biomass components decomposition as hemicellulose > cellulose >>> lignin. Based on different reasoning, Yang et al. (2007) proposed different decomposition regions of 220-300 °C, 300-340 °C and >340 °C for hemicellulose, cellulose and lignin, respectively. Lignin is the last to decompose due to its heavy cross linked structure (Guo & Lua, 2001).

Several techniques are available to study the kinetics of biomass decomposition. Among these, thermogravimetric analysis (TGA) is the most popular and simplest technique (Luangkiattikhun et al., 2008), based on the observation of sample mass loss against time or temperature at a specific heating rate. TGA provides high precision (Várhegyi et al., 2009), fast rate data collection and high repeatability (Yang et al., 2004) under well defined kinetic control region.

Very few attempts have been carried out to study the kinetics of empty fruit bunch (EFB) and palm shell (PS) using TGA. Guo & Lua (2001) presented the effect of sample particle size and heating rate on pyrolysis process and kinetic parameters for PS. They concluded a first order reaction mechanism for the decomposition of PS at different heating rates. They also suggested higher heating rates for faster and easy thermal decomposition of PS. Yang et al. (2004) studied activation energy for decompositions of hemicellulose and cellulose in EFB and PS by considering different temperature region for first order kinetic reaction. They evaluated average activation energy and pre-exponential factor from single-step decompositions of hemicellulose and cellulose. Luangkiattikhun et al. (2008) considered the effect of heating rate and sample particle size on the thermogram behaviour and kinetic parameters for palm oil shell, fibre and kernel. They observed that there is no significant effect of particle size on the thermogram behaviour at lower temperature i.e. <320 °C for palm oil shell. They further proposed nth order reaction mechanism to evaluate the kinetic parameters based on different models.

Previous works reported on EFB and PS kinetics were based on single heating rate in which activation energy is only a function of temperature. The present work evaluate the kinetic parameters based on a method, which requires at least three sets of experimental data generated at different heating rates. This method allows the dependence of activation energy on temperature and conversion at a desired heating rate (Vyazovkin & Wight, 1999) Secondly, lignin decomposition in EFB and PS is not intensively studied at relatively high heating rates. Present work considers lignin decomposition in EFB and PS to understand the effect of lignin content on kinetic parameters and decomposition rate. Furthermore, pure lignin decomposition is studied based on its thermogram analysis and kinetic parameters.

In this work, the kinetics of biomass decomposition which includes EFB, PS, pure cellulose and lignin were investigated using TGA under non-isothermal conditions. The detail thermogram analysis was presented to understand the decomposition of cellulose, hemicellulose and lignin as major components in lignocellulosic biomass. The decomposition kinetics of cellulose and lignin were studied under single-step first order kinetic model. Meanwhile, the decomposition of EFB and PS were reported based on single-step nth order kinetic model. Activation energy, pre-exponential factor and order of reaction were determined and discussed in comparison to the values reported in the literature.

2. MATERIALS AND METHODS

2.1. Materials Preparation and Experimental Procedure

Cellulose in fibrous powder form and lignin in brown alkali powder form were purchased from Sigma Aldrich Sdn. Bhd., Malaysia. EFB and PS were collected from local palm oil industry in Perak, Malaysia. Biomass samples were dried at 105°C and the weighted was monitored at one hr interval, until the readings became constant. Samples were then grinded to particle size of 150-250μm. The method for drying, characterization and analysis were given in previous work (Abdullah et al., 2010). The biomass, pure cellulose and lignin properties are given in Tables 1 and 2.

The biomass decomposition experiments were carried out in EXSTAR TG/DTA 6300 (SII, Japan). $N_2$was used as inert gas with a constant flow rate of 100 ml/min for the entire range of experiments. The sample initial weight used in all experiments was within the range of 3-6 mg. TG experiments were performed at heating rate of 10, 30 and 50 °C/min. All samples were first heated from 50 °C to 150 °C where it was kept constant for 10 min to remove moisture content, and then heated up to the final temperature of 800 °C. All experiments were carried twice for reproducibility. No significant variations were observed in the second experimental measurements.

2.2. Kinetic Parameters Determination

The biomass decomposition rate under non-isothermal condition is described (Cai & Bi, 2009).

$$\frac{d\alpha}{dT} = \frac{A}{\beta}\exp(-\frac{E}{RT})f(\alpha) \tag{1}$$

$f(\alpha)$ depends on the reaction mechanism as listed in Table 3 and α is the mass fraction reacted.

$$\alpha = \frac{w_0 - w}{w_0 - w_f} \tag{2}$$

Where, 0 and f shows initial and final sample weight

$$g(\alpha) = \int_0^{\alpha} \frac{d\alpha}{f(\alpha)} = \frac{A}{\beta}\int_0^{T} \exp(-\frac{E}{RT})dT = \frac{AE}{\beta R}p(\frac{E}{RT}) \tag{3}$$

p(E/RT) function has no exact analytical solution, and therefore different approximations are reported to evaluate the function (Budrugeac et al., 2000). The method developed by Flynn-Wall (Flynn & Wall, 1966), Ozawa (1965) using Doyle's approximation (1961) is the most popular and commonly used by several researchers for biomass decomposition (Cai & Bi, 2009; Hu et al., 2007; Zhouling et al., 2009).

$$p(\frac{E}{RT}) = 0.0048e^{-1.0516(\frac{E}{RT})} \tag{4}$$

(3) is then rearranged for β

$$ln\beta_i = ln(\frac{A_\alpha E_\alpha}{Rg(\alpha)}) - 5.331 - 1.0516\frac{E_\alpha}{RT_{\alpha,i}} \tag{5}$$

Where; w = sample weight (mg); β = heating rate (K/min); R = universal gas constant (8.314×10^{-3} kJ mol^{-1} K^{-1})

To determine the activation energy, ln β_i vs. $1/T_{\alpha,i}$ is plotted for different α values and heating rates (i) to give a straight line and the slope of which gives the activation energy (Doyle, 1961; Ozawa, 1965;Zsakó, & Zsakó, 1980; Flynn & Wall, 1966).

TABLE 1. Biomass analysis (wt% dry basis)

Analysis	EFB	PS
Volatiles	84.61	81.03
Ash	5.50	4.10
Fixed Carbon (by difference)	9.89	14.87
C	40.73	49.65
H	5.75	6.13
N	1.40	0.41
S	0.22	0.48
O (by difference)	51.90	43.33
Cellulose[a]	38.30	20.8
Hemicellulose [a]	35.30	22.7
Lignin[a]	22.10	50.7

[a] Kelly-Yong et al. (2007)

TABLE 2. Pure cellulose and Lignin properties (wt % dry ash free)

Analysis	Cellulose	Lignin
C	43.09	47.71
H	5.96	4.53
N	0.13	0.04
S	0.14	4.24
O (by difference)	50.67	43.40

TABLE 3. Different $f(\alpha)$ and $g(\alpha)$ values based on kinetic control regime (Ahmad et al., 2009)

Function	f(α)	g(α)
First order reaction	$1-\alpha$	$-\ln(1-\alpha)$
Second order reaction	$(1-\alpha)^2$	$(1-\alpha)^{-1}-1$
Third order reaction	$1/2(1-\alpha)^3$	$(1-\alpha)^{-2}-1$
nth order reaction	$(1-\alpha)$	$1-(1-\alpha)^{1-n}/1-n$

2.3. Model for kinetic parameter determination

The following assumptions are considered for the decomposition of EFB, PS, pure cellulose and lignin.

- Reaction is purely kinetic controlled.
- The decompositions follow single-step processes.
- First order reaction kinetics is considered for pure cellulose and lignin and PS and EFB kinetics are assumed to be nth order.
- No secondary reaction takes place among the gaseous products.

3. RESULTS AND DISCUSSIONS

3.1. Thermogram Analysis

The TG and DTG curves for cellulose, lignin, EFB and PS at different heating rates are shown inFigures 1-4. The effect of different heating rate can be described by a lateral shift appeared at high heating rates. These lateral shifts are due to the thermal lag effect between surrounding and biomass particles (Yang et al., 2004; Luangkiattikhun et al., 2008). As a result, conversions are delayed at high heating rates. Thermal lag effect is due to the small heat conductive property of biomass particles (Zhang et al., 2006).

In the DTG curves (Fig. 1-4, b) for all samples, high decomposition rate was observed at 50 °C/min, which shows the increase of thermal decomposition rate of biomass at high heating rates.

The investigated EFB exhibited the decomposition rate corresponds to -41 wt%/min which is higher than -33 wt%/min of PS at 50 °C/min (see Fig. 5). The high decomposition rate for EFB and PS appeared at 342 and 382 °C, respectively. It is important to consider that 60 wt% of EFB and PS is decomposed at 400 and 429 °C for 50 °C/min. These results depict relatively easy and fast decomposition for EFB as compared to PS. This fast decomposition of EFB may be attributed to the comparatively high volatiles matter and low lignin content present in EFB as compared to PS. Conversely, pure cellulose and lignin decomposition rate is the highest and lowest among all species which is -124 and -19 wt%/min at 50 °C/min, respectively. Furthermore, the highest decomposition rate for cellulose and lignin is observed at 386 and 418 °C.

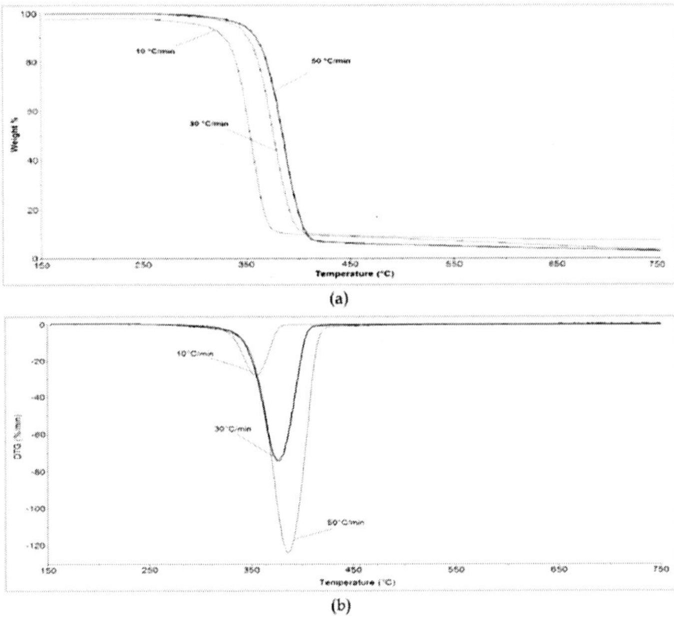

FIGURE 1. Cellulose (a) TG and (b) DTG curves

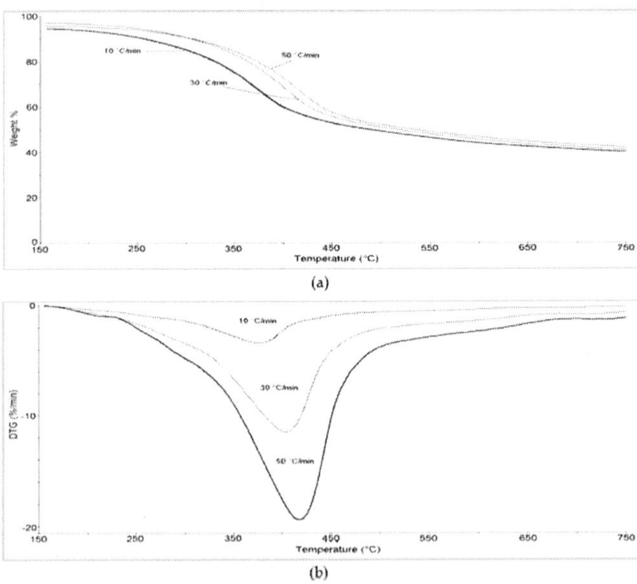

FIGURE 2. Lignin (a) TG and (b) DTG curves

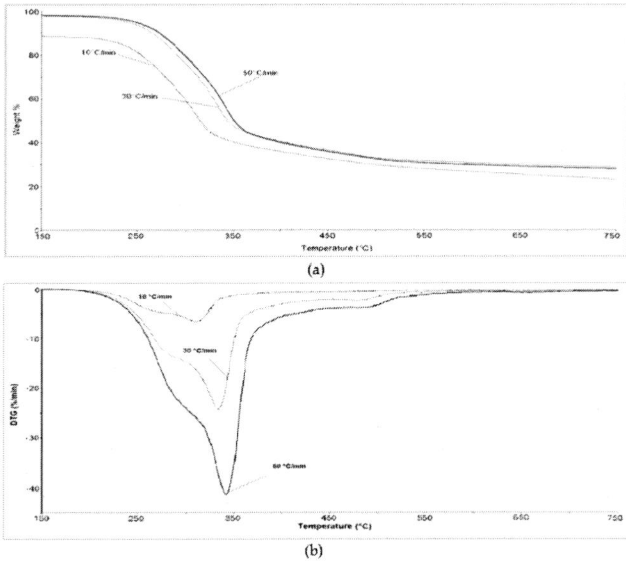

FIGURE 3. EFB (a) TG and (b) DTG curves

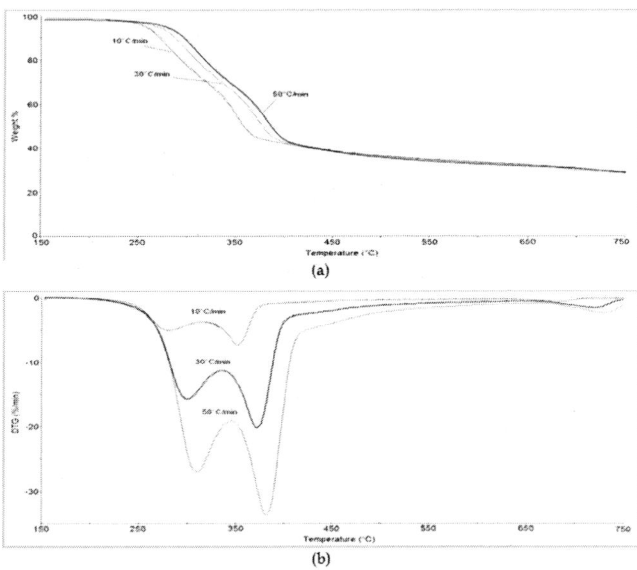

FIGURE 4. PS (a) TG and (b) DTG curves

The TG and DTG curves for EFB and PS are given in Figures 3-4. In these figures, the first peak represents the decomposition of hemicellulose. The second peak, which is sharper, gives the highest rate corresponds to the cellulose decomposition. The decomposition range of hemicellulose and cellulose of EFB is between 240-300 °C and 300-340 °C, respectively, at heating rate of 10 °C/min. Decomposition rate of hemicellulose in PS falls almost in the same temperature region as for EFB but higher decomposition range for cellulose (340-370 °C). It is important to consider that the cellulose decomposition rate in PS is in the same temperature region as pure cellulose (340-370 °C at 10 °C/min). The tail at high temperature shows lignin decomposition as found by Yang et al. (2004) and Luang kiattikhun et al. (2008). In the present study, at 10 °C/min, no lignin decomposition was observed for EFB and PS. Similar observation is reported by Yang et al. (2004) for heating rate of 10 °C/min at temperature >340 °C. At higher heating rates, there is some small lignin decomposition observed for EFB and PS which is in the range of 450-530 °C and 680-750 °C, respectively. Different region for lignin decomposition in EFB and PS may be due to different lignin structure and composition in both species.

Among all species, lignin decomposition produced highest residual fraction of ~40% followed by ~27% of EFB and PS and <7% for cellulose, respectively. High residual fraction for lignin shows its high resistance to thermal decomposition which can be seen by its lowest decomposition rate.

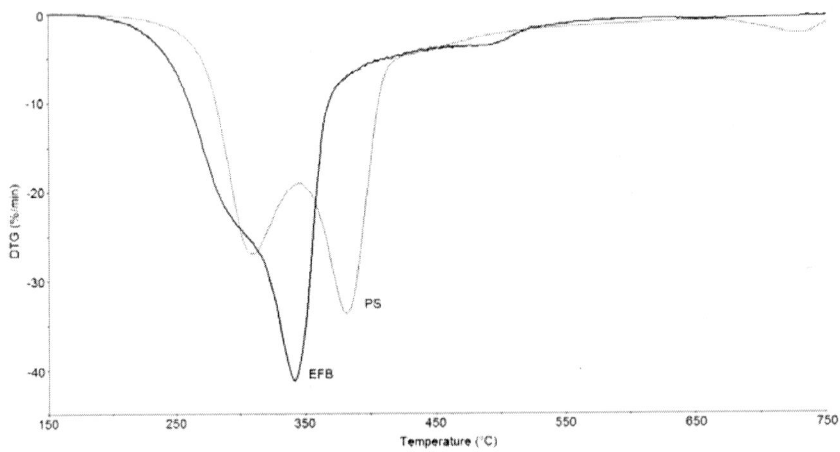

FIGURE 5. EFB and PS DTG curves at 50 °C/min

3.2. Kinetic Parameters

Kinetic parameters are evaluated using equation (5) at a heating rate of 10, 30 and 50 °C/min. Figures 6-9 shows the plot of $\ln\beta_i$ against 1/T at different mass fraction reacted (α) for cellulose, lignin, EFB and PS. The higher $\ln\beta_i$ values shows high heating rate (50 °C/min) followed by 30 and 10 °C/min. Kinetic parameters evaluated at each α are given in the Tables 4-7. For pure cellulose and lignin, the kinetic parameters are determined at α=0.1 to 0.8 and α=0.1 to 0.6, respectively. EFB and PS kinetic parameters are evaluated at α=0.1 to 0.7. Among all samples, pure lignin produced highest residual fraction and hence kinetic parameter determined up to α=0.6. The correlation coefficients (R^2) determined are higher than 0.991 for all cases.

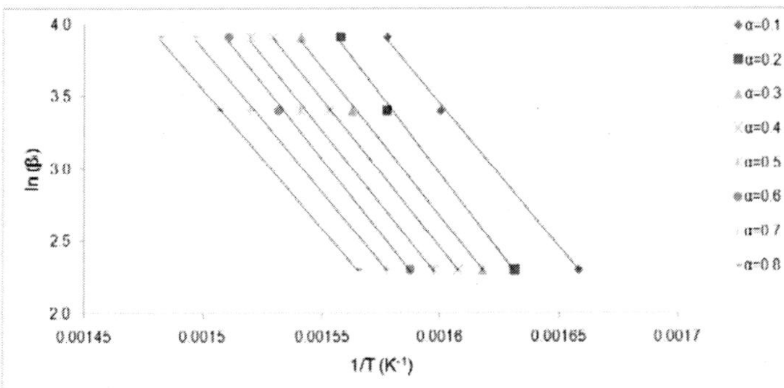

FIGURE 6. The dependence of $\ln(\beta_i)$ on 1/T at a different α values of cellulose (solid lines shows linear fitting)

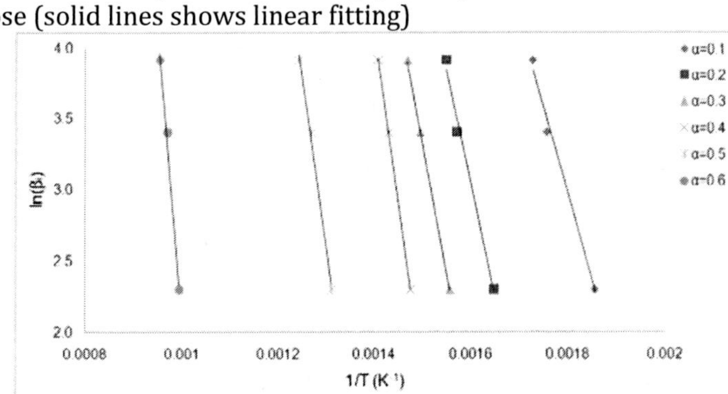

FIGURE 7. The dependence of $\ln(\beta i)$ on 1/T at a different α values of lignin (solid lines shows linear fitting)

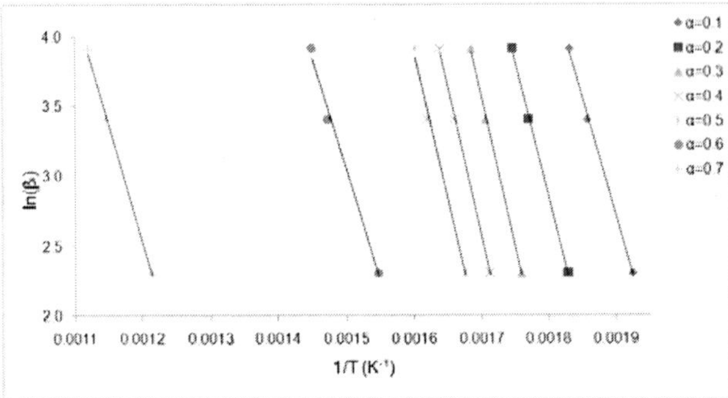

FIGURE 8. The dependence of ln(βi) on 1/T at a different α values of EFB (solid lines shows linear fitting)

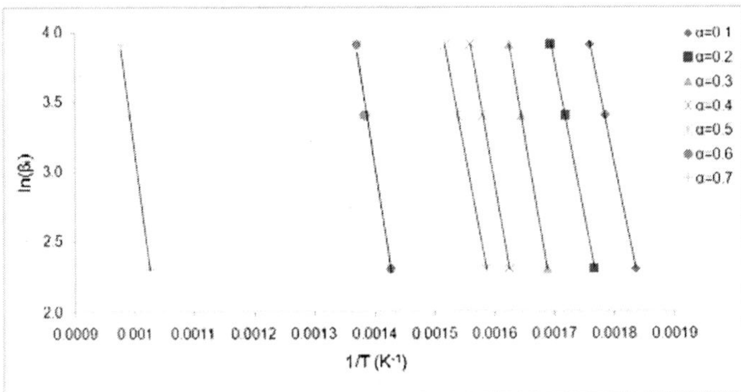

FIGURE 9. The dependence of ln(βi) on 1/T at a different α values of EFB (solid lines shows linear fitting)

Table 4. Kinetic parameters of cellulose at different α values

α	E (kJ/mol)	R^2	A (m^{-1})
0.1	155.31	0.996	3.9×10^{13}
0.2	170.14	0.996	3.9×10^{13}
0.3	162.95	0.998	1.1×10^{13}
0.4	161.12	0.999	9.2×10^{12}
0.5	161.64	0.997	1.1×10^{13}
0.6	163.66	0.997	1.8×10^{13}
0.7	157.03	0.998	5.3×10^{12}
0.8	151.53	0.991	1.9×10^{13}
Average	160	0.997	1.2×10^{13}

TABLE 5. Kinetic parameters of lignin at different α values

α	E (kJ/mol)	R^2	A (m^{-1})
0.1	95.01	0.990	9.1E+10^7
0.2	123.28	0.990	4.4E+10^9
0.3	143.67	0.999	7.9E+10^{10}
0.4	183.89	0.999	3.8E+10^{13}
0.5	184.22	0.995	1.2E+10^{12}
0.6	318.48	0.992	1.2E+10^{16}
Average	175	0.994	2.06E+10^{15}

TABLE 6. Kinetic parameters of EFB at different α values

α	E (kJ/mol)	R^2	A (m^{-1})
0.1	135.88	0.998	3.9×10^{13}
0.2	154.59	0.996	6.0×10^{14}
0.3	169.01	0.998	4.5×10^{15}
0.4	171.96	0.998	4.7×10^{15}
0.5	169.92	0.993	9.1×10^{13}
0.6	125.60	0.997	9.4×10^{10}
0.7	131.69	0.997	4.0×10^{9}
Average	151	0.997	1.4×10^{15}

TABLE 7. Kinetic parameters of PS at different α values

α	E (KJ/mol)	R^2	A (m^{-1})
0.1	165.20	0.999	6.5×10^{15}
0.2	176.23	0.999	1.9×10^{16}
0.3	200.13	0.999	6.2×10^{17}
0.4	194.15	0.998	5.4×10^{16}
0.5	186.27	0.999	7.6×10^{15}
0.6	214.97	0.992	5.9×10^{16}
0.7	255.68	0.998	4.0×10^{14}
Average	199	0.998	1.1×10^{17}

The kinetic parameters for EFB, PS, pure cellulose and lignin are determined as listed in the Table 8and compared with experimental works reported. The average activation energy of cellulose is 160 kJ/mol and pre-exponential factor is 1.2×10^{13} m^{-1} for first order kinetic model. Several researchers found that first order kinetics fit well for cellulose decomposition reaction (Várhegyi et al., 1997;Gronli et al., 1999; Hu et al., 2007).

Activation energy evaluated for cellulose in the present study is in good agreement with the work reported by Zhang et al. (2009) for a first order kinetic model. Nevertheless, the value is comparatively low corresponding to values reported by Varhegyi et al. (1997), Gronli et al. (1999) and Yang et al. (2004). The reason may be different source of cellulose and different method followed by the authors. Secondly, this may be due to relatively high heating rates used in the present study. Meanwhile, Hu et al. (2007) reported high activation energy (233 kJ/mol) using Flynn-Wall-Ozawa method at low heating rates of 2.5, 5, 10 °C/min. Gronli et al. (1999) observed the effect of different heating rates on activation energy for cellulose and found low activation energy at high heating rates. Similarly, much lower activation energies were found by Milosavljevic et al. (1995) at high heating rates. The same effect was also observed for pre-exponential factor. This is due to the heat transfer limitation between sample particles and the surroundings at high heating rates. Varhegyi et al. (1997) suggested utilization of low sample mass to minimize the heat transfer limitations at high heating rates.

TABLE 8. Comparison of kinetic parameters

Biomass	Kinetic Parameters			R^2	Reference
	E (kJ/mol)	A (m^{-1})	n (-)		
Cellulose	160	1.2×10^{13}	1	0.997	This study
	175	7.16×10^{12}	1	0.999	Zhang et al. (2009)
Lignin	175	2.06×10^{15}	1	0.994	This study
	171	2.73×10^{25}	1	-	Morugun et al. (2008)
EFB	151	1.4×10^{15}	5.3	0.997	This study
	61	3.14×10^{2}	1	0.991	Yang et al. (2004)
PS	199	1.1×10^{17}	5.0	0.998	This study
	111	5.27×10^{7}	2.54	-	Luangkiattikhun et al. (2008)

Pure lignin is decomposed with average activation energy of 175 kJ/mol and pre-exponential factor of 2.06×10^{15} m^{-1} for first order kinetics model. These kinetic parameters are in good agreement with work reported by Murugan et al. (2008).

EFB and PS are decomposed with the average activation energy of 151 and 199 kJ/mol and reaction order of 5 and 5.3, respectively. The corresponding average pre-exponential factors evaluated for EFB and PS are 1.4×10^{15} and 1.1×10^{17} m^{-1}, respectively. These values are somehow larger than those reported by Yang et al. (2004) for first order reaction kinetics. Luangkiattikhun et al. (2008) and Guo & Lua, (2001) observed lower activation energy and pre-exponential factor based on single-step nth order and first order kinetic model for PS. However, in these studies, comparatively high values were obtained using two step kinetic models.

Based on kinetic parameters, it is easy to decomposed EFB as compared to PS, pure cellulose and lignin. The order of decomposition from fast to slow is EFB > cellulose > lignin > PS.

4. CONCLUSION

A biomass decomposition study has been carried out to investigate different breakdown region and kinetic parameter evaluation for EFB and PS. As major components of biomass, pure cellulose and lignin decomposition kinetics were also studied. TG and DTG curves were studied in detail to understand the major decomposition region in EFB, PS, pure cellulose and lignin. The kinetic parameters of EFB and PS are found to be higher compared to reported values in the literatures. This difference may be due to the different methods for kinetic parameter determination and relatively high heating rates used in the present study. Based on the kinetic parameters, PS was difficult to decompose as compared to EFB. The possible reason is may be relatively high lignin content present in PS. This high lignin content was also responsible for low decomposition rate of PS as compared to EFB. Pure lignin had the lowest decomposition rate among all the species. Moreover, lignin content in PS was decomposed at high temperature as compared to EFB based on higher heating rates.

5. ACKNOWLEDGEMENTS

The author thanks the Petroleum Research Fund (PRF) of PETRONAS and Universiti Teknologi PETRONS for their financial support.

REFERENCES

1. Abdullah, S.S. & Yusup, S. (2010). Method for Screening of Malaysian Biomass Based on Aggregated Matrix for Hydrogen Production through Gasification. Journal of Applied Sciences, 1812-5654 24 10 3301 3306
2. Ahmed, A. K. Gupta, 2009 Hydrogen Production from Polystyrene Pyrolysis and Gasification: Characteristics and Kinetics International Journal of Hydrogen Energy 34 15 6253 6464 1520-5029
3. P. Budrugeac, D. Homentcovschi, E. Segal, 2000 Critical Analysis of the Isoconversional Methods for Evaluating the Activation Energy. I. Theoretical Background Journal of Thermal Analysis and Calorimetry 63 2 457 463 1388-6150
4. J. Cai, L. Bi, 2009 Kinetic Analysis of Wheat Straw Pyrolysis using Isoconversional Methods Journal of Thermal Analysis and Calorimetry 98 1 325 330 1388-6150
5. Demirbas, 2001 Biomass Resource Facilities and Biomass Conversion Processing for Fuels and Chemicals Energy Conversion and Management 42 11 1357 1378 0196-8904
6. D. Doyle, 1961 Kinetic Analysis of Thermogravimetric Data Journal of Applied Polymer Science 5 15 285 292 0021-8995
7. J. H. Flynn, L. A. Wall, 1966 A Quick, Direct Method for the Determination of Activation Energy fromThermogravimetric Data. Journal of Polymer Science, Part B: Polymer Letter, 4 5 323 328 1542-6254
8. Gronli, M.; Antal, M.J. & Varhegyi, G. (1999). A Round-Robin Study of Cellulose Pyrolysis Kinetics by Thermogravimetry. Industrial Engineering and Chemistry Research, 1520-5045 6 38 2238 2244
9. J. Guo, A. C. Lua, 2001 Kinetic Study on Pyrolytic Process of Oil-Palm Solid Waste Using Two-Step Consecutive Reaction Model Biomass and Bioenergy 20 3 223 233 0961-9534
10. Hu, S.; Jess, A. & Xu, M. (2007). Kinetic Study of Chinese Biomass Slow Pyrolysis: Comparison of Different Kinetic Models. Fuel, 0016-2361 17-18 86 2778 2788
11. T. L. Kelly-Yong, K. T. Lee, A. R. Mohamed, S. Bhatia, 2007 Potential of Hydrogen from Oil Palm Biomass as a Source of Renewable Energy Worldwide Energy Policy 35 11 5692 5701 0301-4215

12. Khan, Z.; Yusup, S.; Ahmad, M.M.; Chok, V.S.; Uemura, Y. & Sabil, K.M. (2010). Review on Hydrogen Production Technologies in Malaysia. International Journal of Engineering & Technology, 2077-1185 02 10 111 118
13. P. Luangkiattikhun, C. Tangsathitkulchai, M. Tangsathitkulchai, 2008 Non-isothermal Thermogravimetric Analysis of Oil-Palm Solid Wastes Bioresource Technology 99 5 986 997 0960-8524.
14. Milosavljevic, E. M. Suuberg, 1995 Cellulose Thermal Decomposition Kinetics: Global Mass Loss Kinetics Industrial Engineering and Chemistry Research, 34 4 1081 1091 1520-5045
15. P. Murugan, N. Mahinpey, K. E. Johnson, M. Wilson, 2008 Kinetics of the Pyrolysis of Lignin Using Thermogravimetric and Differential Scanning Calorimetry Methods Energy & Fuels 22No. 2720 2724 0887-0624
16. Ozawa, T. (1965). A New Method of Analyzing Thermogravimetric Data. Bulletin of the Chemical Society of Japan, 0009-2673 11 38 1881 1886
17. S. H. Shuit, K. T. Tan, K. T. Lee, A. H. Kamaruddin, 2009 Oil Palm Biomass as a Sustainable Energy Source: A Malaysian Case Study. Energy, 34 4 1225 1235 0360-5442
18. G. B. Várhegyi, H. Chen, S. Godoy, 2009Thermal Decomposition of Wheat, Oat, Barley, and Brassica carinata Straws. A Kinetic Study. Energy and Fuels, 23 2 646 652 0887-0624
19. Várhegyi, G.; Antal, M. J.; Jakab, E. & Szabó, P. (1997). Kinetic Modeling of Biomass Pyrolysis. Journal of Analytical and Applied Pyrolysis, 0165-2370 1 42 73 87
20. S. Vyazovkin, C. A. Wight, 1999 Model-Free and Model-Fitting Approaches to Kinetic Analysis of Isothermal and Nonisothermal Data Thermochimica Acta, 340-341 53 68 0040-6031
21. Yang, H.; Yan, R.; Chin, T.; Liang, D.T.; Chen, H. & Zheng, C. (2004). Thermogravimetric Analysis–Fourier Transform Infrared Analysis of Palm Oil Waste Pyrolysis. Energy and Fuels, 0887-0624 6 18 1814 1821
22. Yang, H.; Yan, R.; Chen, H.; Lee, D.H. & Zheng, C. (2007). Characteristics of Hemicellulose, Cellulose and Lignin Pyrolysis. Fuel, 0016-2361 86 1781 1788
23. X. Zhang, W. de Jong, F. Preto, 2009 Estimating Kinetic Parameters in TGA using B-spline Smoothing and the Friedman Method Biomass and Bioenergy 33 10 1435 1441 0961-9534

24. X. Zhang, M. Xu, R. Sun, L. Sun, 2006 Study on Biomass Pyrolysis Kinetics Journal of Engineering for Gas Turbines and Power 128 3 493 496 0742-4795
25. Zhouling.; Chang-qing, S.; Chao-yu, L.; En-cheng, J. & Xi-wei, X. (2009). Pyrolysis Characteristics and Kinetics of Cotton Stalk, International Conference on Energy and Environment Technology, 978-0-76953-819-8 768 772
26. J. Zsakó, J. Zsakó, 1980Kinetic Analysis of Thermogravimetric Data XIV. Three integral methods and their computer programs. Journal of Thermal Analysis and Calorimetry, 19 2 333 345 1388-6150

4
CHAPTER

EVALUATION OF RENEWABLE ENERGY VULNERABILITY TO CLIMATE CHANGE IN BRAZIL: A CASE STUDY OF BIOFUELS AND SOLAR ENERGY

Antonio Oscar Jr., Wanderson Luiz Silva, Vera Ruffato, Renata Barreto, Marcos Freitas

Federal University of Rio de Janeiro, Rio de Janeiro, Brazil

ABSTRACT

This study aims to calculate indicators and indexes to subsidize the analysis of vulnerability and adaptation of the renewable energy sector to climate change in Brazil, focusing on biofuels and solar energy. For biofuels, in general, the Brazilian coast will be a propitious area for agricultural productivity during the XXI century, but these are areas historically intended for occupation and development of the urbanization process, that is, with limited land availability and supply for primary production. In some parts of the Northeast, Midwest and South of the country, offer for the cultivation land will be reduced. For the solar energy is observed that Brazil has area and highly expressive power for the use of this power, both today and in the coming decades, especially in the North, Northeast and Midwest. In statistical terms, the Mann-Kendal test and Sen's Bend point to a very weak tendency to useful radiation indicator in all regions of Brazil by 2100. In addition, it is projected a significant increase in mean air temperature by the end of XXI century across the country that can mean a reduction in power conversion capability, which is sensitive to ambient temperature variations, especially in the Midwest and North of the country.

Keywords: Renewable Energy, Climate Change, Brazil, Biofuels, Solar Energy

1. INTRODUCTION

Climate change increasingly occupies an important position within the various sectors of society, both nationally and internationally. Several studies, including the fifth assessment report of the Intergovernmental Panel on Climate Change [1] , indicate that the growing trend of greenhouse gases is the main agent of such changes. Thus, there is an imminent concern about the mitigation of these future climate changes, making it necessary to establish measures to adapt to possible impacts.

As the power generation and consumption are key to socio-economic development, it is essential to understand how the energy production infrastructure responds to the clash of climate change. From this knowledge is possible to measure the vulnerability of a nation and promote development of alternative means of it, offering appropriate prospects for the future. In Brazil, due to its characteristic of primary producer associated with continental proportions of its territory and climatic aspects, its energy matrix were guided primarily in renewable energy.

Energy sources exploited in Brazil include hydropower, thermal power, solar power, biofuels and wind energy. According to the National Electric Energy Agency [2] , the country has 3336 projects of electricity generation in operation, representing an output of approximately 118 Gigawatts. Thus, significant climate changes may represent substantial impacts for development and for the country economy in this sector. Specifically this research, the main focus will be on solar energy and biofuels, as these sources have significant enhancement conditions and vulnerability to present and future climate scenarios.

The energy originated from biofuels is from renewable sources which in Brazil are essentially represented by biodiesel and ethanol. For proper use of land with purposes of exploitation of agricultural potential for the generation of biofuels, in addition to economic and social factors, it must take into account basic environmental factors, considering the physicochemical requirements of each culture, as well as the ecology of the region. Among the physical and ecological conditions, the climate plays an important role, especially with regard to water availability, which depends directly on rainfall supply and intensity of evapotranspiration process, elements that act as limiter to vegetation.

For its intertropical location, Brazil has a high incidence of solar radiation throughout the year, making this a potential source for the exploration of energy in order to supply the increasing demand of energy in the country. Most of Brazil is located relatively close to the equator, so that there are not wide variations in daily sunlight. However, most of the population and socio-economic activities of the country are concentrated in more distant regions of equator, as in the Southeast and South.

In view of the above perspectives, this work presents a vulnerability analysis and interpretation of the impacts (both positive and negative) of future climate changes in the renewable energy sector of biofuels and solar energy, in order to support researches that aim to adaptations and implementations in the current Brazilian energy matrix.

2. METODOLOGY

2.1. Data from Climate Model

The Numerical Models of the Earth System (NMESs) are the most promising tools to develop projections of future climate change. Through the simulation of important physical and dynamical processes, NMESs can represent the complex nonlinear interactions that influence the climate as well as the interactions between the components of the Earth system (mainly atmosphere, biosphere and hydrosphere) and feedback mechanisms, including changes in the frequency and intensity of extreme events [3] . In addition, NMESs can simulate future climates in response to changes in the concentration of greenhouse gases and aerosols.

For this study, the regional climate was simulated by Eta model of Weather Forecasting and Climate Studies Center/National Institute for Space Research (CPTEC/INPE), which is coming from the original Eta model [4] developed at the University of Belgrade (Serbia) and operationally implemented by the National Center for Environmental Predictions (NCEP) in the United States. The Eta 20 km regional model scenarios were simulated using the global model boundary conditions MIROC (Model for Interdisciplinary Research on Climate), a model developed in cooperation with the climate research center in Japan. From these integrations, the model is called Eta-Miroc5 [5] [6] .

In this study, the average of thirty years is used as reference model for the processing of data and evaluation of future scenarios of the indicators described hereafter. The RCP 8.5 scenario [1] was selected to perform the climate change analyzes. Historical values (1961-1990) and projected future values (2011-2040, 2041-2070 and 2071-2100) were calculated. Then the future-present differences were computed in order to verify the simulations of future trends.

2.2. Biofuels Indicator

Assuming that water availability is fundamental to the farm of genres used for the production of biofuels, including ethanol, it is used the aridity index (D) proposed by [7]:

$$D = \frac{1}{(0.01 * IM) + 1}$$

where $IM = 100 * \left(\frac{PRECP}{EVTP} - 1\right)$ is the Thorntwaite Moisture Index, PRECP is the precipitation and EVTP is the evapotranspiration.

From this method, it was shown that occurrence of areas of desertification more emphatic process, so those with very low agricultural potential, are in the range $2 < D < 7$.

To complement this methodology is also used as reference humidity data in the soil in 0.1 m, which aims to support the assessment of water availability to the root system of plants. Furthermore, air temperature is also analyzed, from which proposes an index based on its variation (projected temperature-historical temperature), allowing estimation of the thermal stress of the plant in climate change scenarios, fundamental factor that directly affects the photosynthetic capacity of the plant.

2.3. Solar Energy Indicator

Assuming that the use of solar energy depends on the incident solar radiation in a given area, an efficient method of analyzing the view of the offer of this source is the radiation balance in a given region, which can be determined by calculating the balance of radiation. This radiation (Rn) is the available energy in the earth-atmos- phere system and is defined as the radiation balance of all radiative fluxes leading to and from a surface [8] . Therefore, to define a useful radiation indicator (Ru) is required equating the variables of shortwave radiation incident to the surface (OCIS), albedo (α), short wave radiation emerging at the top of the atmosphere (ROCE) and short wave radiation emerging to the surface (OCES). Thus:

$$\mathrm{Ru} = \mathrm{OCIS} + \mathrm{ROCE} - \alpha \cdot \mathrm{OCIS} - \mathrm{OCES}\ \left[\mathrm{W/m^2}\right]$$

The albedo (fraction of incident radiation which is reflected) regarded was the planet mean value, so 0.30 or 30%. The maps show the average annual values of the indicator for a particular day of the year in Brazil, so Ruis multiplied by 24 hours, having a send unit W·h/m^2·day.

The equipment controllers of solar energy can be markedly influenced by variations in ambient temperature. Thus, also it is performed an analysis of projections of future increase or decrease in air temperature along the XXI century in Brazil.

2.4. Statistical Tests

For the evaluation of significance in climatic trends, it is used the Mann-Kendall nonparametric statistical test [9] . This test has been the most appropriate method to analyze the significance of possible climate change in time series [10].

Sen's Bendis also calculated in this work. This is a non-parametric method (assuming a linear trend) used to estimate the trends magnitude [11]. As Sen's Bend is insensitive to outliers and missing data, such testis more rigorous than the usual curvature of regression and provides a more realistic measure of trends in a time series.

3. RESULTS

3.1. Air Temperature

Figure 1 shows the differences in air temperature between the future climate (2011-2040, 2041-2070 and 2071- 2100) and the historical period (1961-1990) projected to Brazil through Eta-Miroc5 model in IPCC RCP 8.5 scenario. The estimates show an increase of about +2.0°C in Brazil during

2011-2040. During 2041-2070, the projections indicate a similar temperature rise, but more sharply in the center of the country, with values reaching up to +3.0°C in the Midwest region and vicinity. Analyzing the period 2071-2100, it is noted that the increase in air temperature is emphatic in Brazil. The simulations show an increase between +2.0°C and +3.0°C in the South and in the east coast, while in up-country the difference can reach up to +5.5°C in the Midwest and something around +4.0°C in the North and in the most part of the Northeast and Southeast regions.

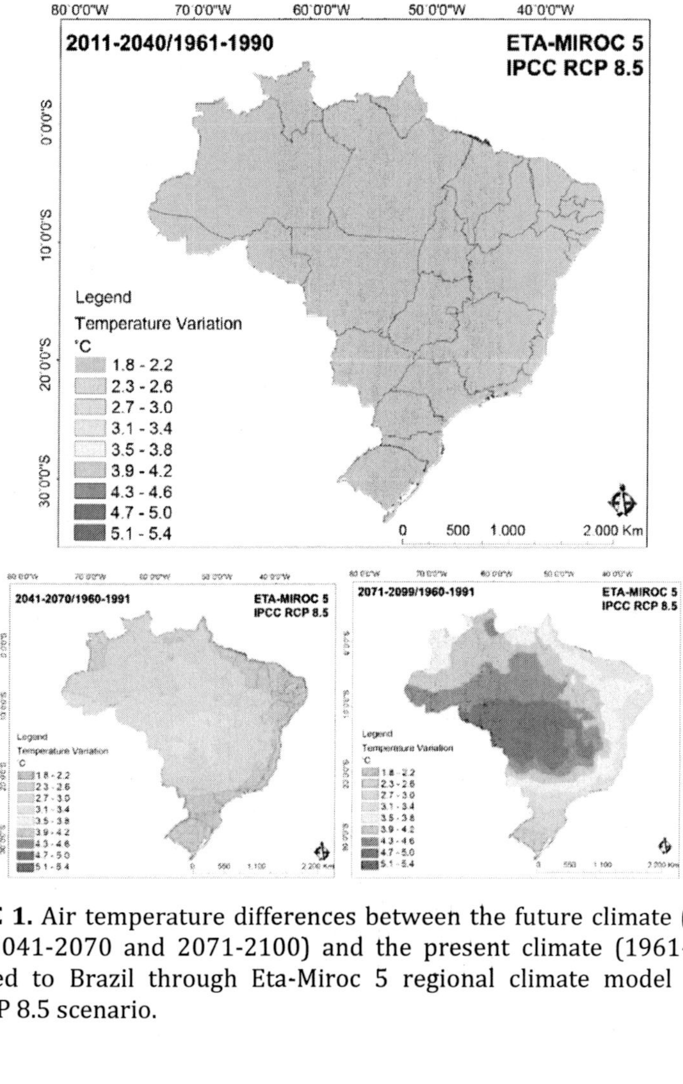

FIGURE 1. Air temperature differences between the future climate (2011-2040, 2041-2070 and 2071-2100) and the present climate (1961-1990) projected to Brazil through Eta-Miroc 5 regional climate model in the IPCCRCP 8.5 scenario.

In statistical terms, the analyzes indicate a significant upward trend in the average air temperature at 2 m for the five Brazilian regions by the end of the century, confirming the possibility of warming and frequency and intensity growth of temperature extreme events (Table 1). Based on this analysis, the Midwest and North Regions Brazil present the highest rates of average temperature rise, with +1.51°C by slice of time and +1.37°C by slice of time, respectively. The region with the lowest magnitude of increase in the period analyzed is the South, with an increase of approximately +0.87°C by slice of time.

3.2. Biofuels (Bioethanol)

The aridity index, formulated in the previous session, is an excellent indicator to assess the water availability in a particular climate type, pointing regions susceptible to the occurrence of severe droughts and desertification process. In the Brazilian, case is notoriously the prone to dryness of the Wild Northeast. From the proposed index (Figure 2), the northern state of Bahia is presented as area of desertification process, which evolves in terms of extension from the time slice 1961-1990 to 2071-2100. It also arise desertification outbreaks in the State of Pará.

The coast of Brazil, both in its east coast as northeast remains with high moisture index, as well as the southeast of the country, which suffers with significant retraction of this area in 2011-2040 period, particularly in the States of São Paulo and Paraná, but from 2041 onwards, it rises again. Certainly the Brazilian coast characteristics associated with a closed topography with high solar radiation and temperature of the surface ensures considerable moisture content and low static stability overlying the air, allowing when the synoptic situation is favorable, the occurrence of convective instability and hence surplus moisture in the environment.

TABLE 1. Statistical analysis of air temperature at 2 meters for Brazilian regions. Values in red were considered significant at a confidence level of 90%.

Region	Mann-Kendall	Sen's Bend
North	0.09	+1.37°C (by slice)
Northeast	0.09	+1.22°C (by slice)
Midwest	0.09	+1.51°C (by slice)
Southeast	0.09	+1.27°C (by slice)
South	0.09	+0.87°C (by slice)

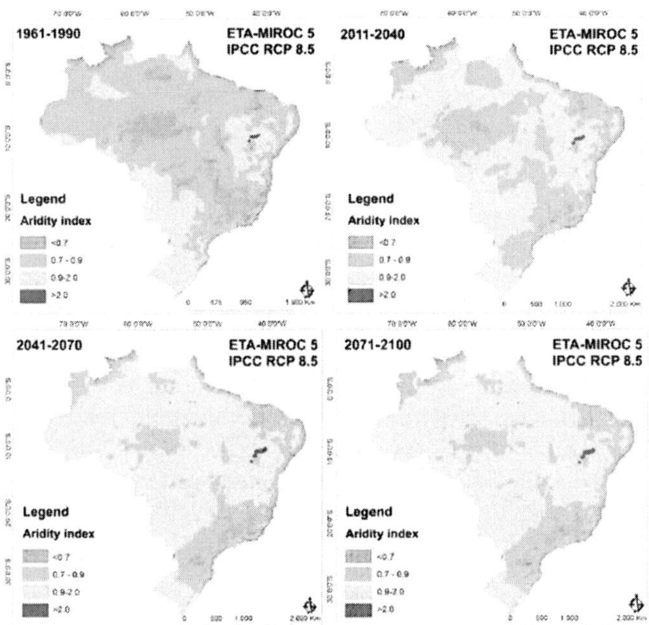

FIGURE 2. Present climate (1961-1990) and simulated future climate (2011-2040, 2041-2070 and 2071-2100) projected for the aridity index in Brazil through Eta-Miroc5 regional climate model in the IPCCRCP 8.5 scenario.

The most significant impact will be felt in the Central West and North of Brazil, where there is significant reduction in available moisture over the time slices, especially in the Amazon and Tocantins, Mato Grosso and Goiás, with increased susceptibility to dryness situations.

In quantitative terms (Figure 3), the projection data of the IPCC RCP 8.5 scenario from Eta-Miroc5 show that in the period 1961-1990, over 74% of Brazil had aridity index below 0.9, so moisture supply code. However, modeling shows a reversal of this scenario by 2100, where approximately 67% will be in the next dryness class, that is, between the indexes 0.9 and 2.0. The areas in desertification, north of Bahia and in specific situations of Pará, will suffer a 0.2% increase in area extension until the end of the twenty-first century.

As for the statistical analysis, the Mann-Kendall test indicates a significant trend in the level of 90% of the aridity index for Brazilian regions, except for the Southeast (Table 2). The magnitude of this change will be greater in the North of Brazil, which several studies have pointed to the savannization trend of lush rainforest, which corroborates the results. Jarring the other regions, the South has an increased trend of humidity, that is, over the slices, the trend observed in this region is greater availability of moisture.

Soil moisture data (Figure 4) also confirm the spatial trend in the aridity index, so point to an expansion of the water deficit area, especially in the States of Bahia, Piauí and Minas Gerais. The data also show an available surface moisture reduction for the Midwest Region of Brazil and project a little significant change to the Brazilian coast and the Southeast. However, the State of Mato Grosso do Suland Amazon region have significant loss of soil surface moisture over the time slices, common characteristics by virtue of the types of soils developed in these areas.

These data are of great importance, since they are indicative of water availability to the root system of the plant, which has a direct impact on nutrient absorption capacity of it. Thus, from the modeling data is possible to estimate a reduction of the area with water availability for vegetation development, leading directly to the productivity of various genres that are distributed between the Northeast and Midwest regions of Brazil.

Statistical analyzes show a high significance for the Northeast region of the country, whose magnitude is −0.01% reduction in moisture between the slices of time on study (Table 3). The Region that clashes of this feature is the South of Brazil, whose tendency is soil moisture increases of +0.01% between the slices of time, strong by connecting with the results presented to the aridity index in the same area.

TABLE 2. Statistical analysis of the aridity index for Brazilian regions. Values in red were considered significant at a confidence level of 90%.

Region	Mann-Kendall	Sen's Bend
North	0.09	+0.04 (by slice)
Northeast	0.09	+0.02 (by slice)
Midwest	0.09	+0.03 (by slice)
Southeast	0.73	+0.01 (by slice)
South	0.09	−0.03 (by slice)

TABLE 3. Statistical analysis of the soil moisture for Brazilian regions. Values in red were considered significant at a confidence level of 90%.

Region	Mann-Kendall	Sen's Bend
North	0.31	−0.01% (by slice)
Northeast	0.09	−0.01% (by slice)
Midwest	0.31	−0.01% (by slice)
Southeast	0.31	−0.01% (by slice)
South	0.31	+0.01% (by slice)

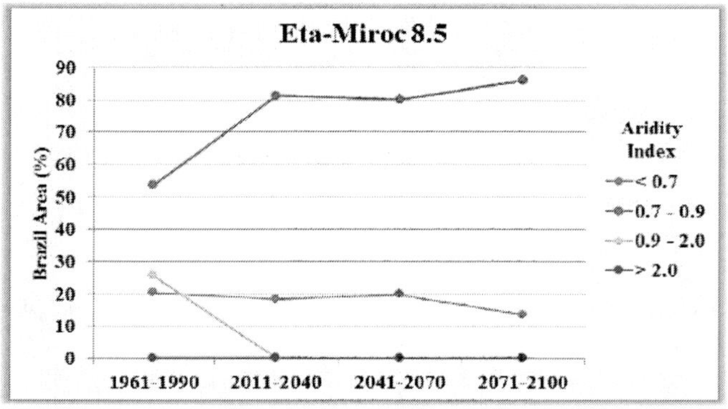

FIGURE 3. Evolution of the areasofaridity index for Brazil through Eta-Miroc5 regional climate model in the IPCCRCP 8.5 scenario.

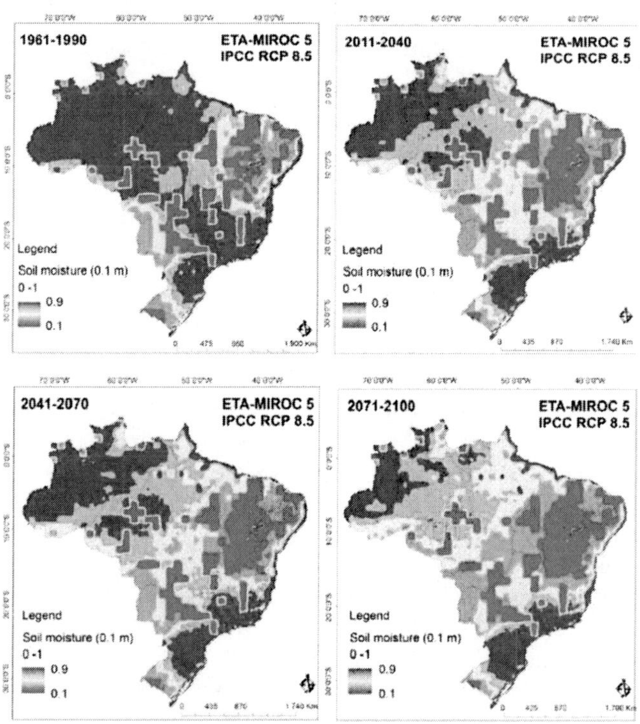

FIGURE 4. Present climate (1961-1990) and simulated future climate (2011-2040, 2041-2070 and 2071-2100) projected forsoil moisture (0.1 m) in Brazil through Eta-Miroc5 regional climate model in the IPCCRCP 8.5 scenario.

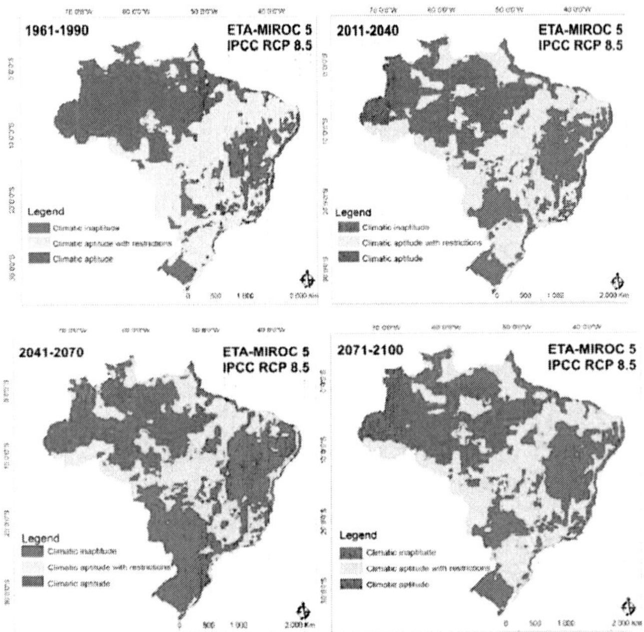

FIGURE 5. Synthesis map of cultivation fitness class to Brazil for present climate (1961-1990) and simulated future climate (2011-2040, 2041-2070 and 2071-2100) through Eta-Miroc5 regional climate model in the IPCCRCP 8.5 scenario.

From the indicators analyzed above, as well as the air temperature (item 3.1), it was possible the development of Figure 5, which is a synthesis map that shows those areas of Brazil with unfitness for cultivation due to projected climate features (areas in red) as well as those areas which have the potential production, but human interventions with the use of more extensive methods of irrigation are required (yellow areas). The map shows also possible situations of ideal suitability for farming, areas where productivity will reach the maximum possible if the other agronomic and environmental conditions prove favorable (green areas). It is noted that by the end of the century, much of the Northeast of Brazil and in the States of Mato Grosso do Sul and Rio Grande do Sul may present climate unsuited to cultivation. However, most of the coast and the North of the country may be able to farming.

Therefore, from the results presented in this section, it is observed that the Southeast region of Brazil, as well as the country's coastline, is favorable areas for agricultural productivity. However, these regions are historically intended for occupation and development of the urbanization process, that is, with limited land availability and supply for production.

3.3. Solar Energy

Analyzing by generalized way all the maps produced by the Eta-Miroc5 regional climate model (Figure 6), there is the same pattern that remains admittedly: an inverse relationship between the intensity of solar radiation and latitude; so at lower latitudes (areas close to the equator, as the North and Northeast of Brazil) solar radiation is high, while in higher latitudes (towards the south of the country), the intensity of this radiation is reduced. This occurs because of the sunlight on the earth's surface, that is, perpendicular in the equatorial and tropical zone and more obliquely toward the poles.

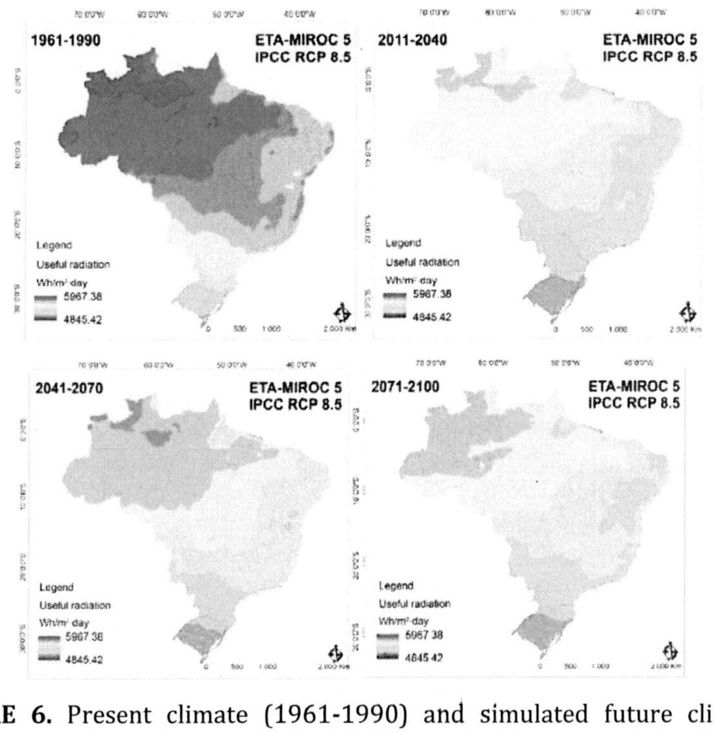

FIGURE 6. Present climate (1961-1990) and simulated future climate (2011-2040, 2041-2070 and 2071-2100) projected for the useful radiation indicator in Brazil through Eta-Miroc5 regional climate model in the IPCCRCP 8.5 scenario.

When the historical period (1961-1990) of the Eta-Miroc5 model is evaluated, it is noted that the Northern region of Brazil has the highest values of useful radiation indicator, and these are established between about 5700 and 5900 W·h/m^2·day. Next, there are the Northeast, Midwest and Southeast, with values around 5500 - 5700 W·h/m^2·day. The southern region has the most marked variation of intensity of this radiation, with values ranging between 5500 W·h/m^2·day in the north area to 4900 W·h/m^2·day in southern part.

As mentioned previously, the eminent values of useful radiation indicator in northern Brazil are due to emphatic solar energy that this area receives during the whole year. Thus, despite the yearly averages for the indicator representation, the values of them remain significantly high. In contrast, the southern region exhibits lower values due to the lower amount of solar energy receipted by this area. In addition, the South has the striking feature of quite often the performance of weather systems that enhance cloud cover, especially the cold fronts.

Observing the future projections of useful radiation indicator generated by the Eta-Miroc5 model forced with the IPCC RCP 8.5 scenario, it appears that the simulated values by the end of the twenty-first century are minor compared with the present climate (historical period). On average, such radiation intensity reductions vary between 4% and 8% in Brazil, values considered not significant, since the country still find within a suitable range for the generation of energy from the short-wave radiation (solar radiation).

Between 2011 and 2040, the North and the northern sector of the Midwest Brazil must have an average value of the useful radiation indicator in about 5600 W·h/m^2·day, so a decrease of around 4% compared with the present climate. The south-center of the Midwest Region, as well as the Northeast and Southeast regions are likely to have values around 5300 W·h/m^2·day, representing thus a reduction about 6% over the historical period. The South should show greater decay in the intensity of solar energy, about 8%, as the average value in this area would be around 5100 W·h/m^2·day.

Accompanying the future projections over 2041 to 2070 periods, there is a negligible increase in the values of useful radiation indicator in Brazil. These elevations are around 2% compared to the previous period (2011-2040), but still remain around 4% lower in relation to the present climate (1961-1990). In the period 2071-2100, the simulated scenario of solar radiation in the country is quite similar to the first discussed future period (2011-2040), with distribution of values of similar manner to that period. There is only one short exception in the States of Minas Gerais and Paraná, which keeps the value of the previous period (2041-2070).

Figure 7 shows the percentage fractions of Brazil area for certain ranges of useful radiation indicator in function of time simulated by the Eta-Miroc5 regional climate model. Taking into consideration a land area of the country at 8.9 million·km^2, currently (1961-1990) and future (2011-2100) the majority (around 61%) has/will have an average annual radiation intensity between 5600 and 6000 W·h/m^2·day. On a slightly lower power range (5200 - 5600 W·h/m^2·day) it is noted and projected to remain around 34% of the total area of Brazil. Finally, about 5% of the country's land have/will have solar radiation between 4800 and 5200 W·h/m^2·day.

Also according to Figure 6 and Figure 7, it is noted that the range of greater intensity of useful radiation indicator (5600 - 6000 W·h/m^2·day) is located on the North and parts of the Northeast, Midwest and Southeast of Brazil. On the other hand, the range with lower solar radiation (4800 - 5200 W·h/m^2·day) is confined in a given stretch of southern Brazil. This reinforces one of the main geographical features and hence climate of Brazil: a country predominantly tropical with strong solar energy available.

In addition, it is projected a significant increase in average air temperature by the end of XXI century across the country, as seen in item 3.1. It means a reduction in energy conversion capacity, sensitive to ambient temperature variations, especially in the North and Midwest of the country. Furthermore, there would be area limitations for the installation of photovoltaic stations due to the large expansion of the urban phenomenon in the national territory, leaving few free areas for these developments.

In statistical terms, the Mann-Kendal test and Sen's Bend point to a few significant trends to useful radiation indicator in all regions of Brazil (0.73), according to Table 4. Probably, this is due a few factors, including: the use of the 30-year average and the low frequency of variation in solar radiation data. However, from the Sen's Bend analysis, it is noted the magnitude of the trend, with unanimity in whimsy decrease in the useful radiation availability, especially in the North and Midwest, which associated with a rise in temperature could interfere in the production ability of energy from photovoltaic systems. However, it is noteworthy that these values are not statistically significant at the 90% level.

Wherefore, it can be concluded that Brazil has area and highly expressive power for using solar energy, both today and in the coming decades, especially in the North, Northeast and Midwest regions. Regarding these considerations, it is worth noting that, since the given values are annual averages in 30 years, it is expected that in the summer the incident solar radiation intensity is more significant, with attenuation only in the presence of cloudiness, mainly in the North, a factor not impeding in a tropical country.

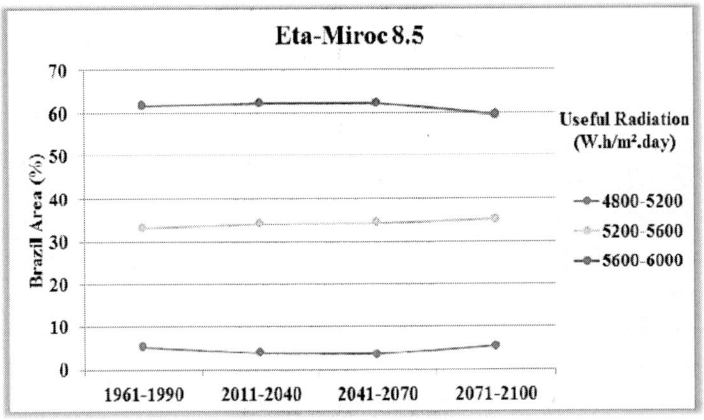

FIGURE 7. Percentage portions of Brazil area to the useful radiation indicator ranges (W·h/m2·day) between 1961 and 2100 simulated by Eta-Miroc5 regional climate model in the IPCCRCP 8.5 scenario.

TABLE 4. Statistical analysis of the useful radiation indicator for Brazilian regions. Values in red were considered significant at a confidence level of 90%.

Region	Mann-Kendall	Sen's Bend
North	0.73	−99.56 W·h/m²·day (by slice)
Northeast	0.73	−84.60 W·h/m²·day (by slice)
Midwest	0.73	−92.91 W·h/m²·day (by slice)
Southeast	0.73	−89.57 W·h/m²·day (by slice)
South	0.73	−71.72 W·h/m²·day (by slice)

4. CONCLUSIONS

The objective of this study was to calculate indicators and indexes so that they subsidized the analysis of vulnerability and adaptation of the renewable energy sector to climate change in Brazil, focusing this research specifically in the sources of biofuels and solar energy.

For biofuels, in general, the Brazilian coast is presented as propitious area for agricultural productivity, but these are areas historically intended for occupation and development of the urbanization process, that is, with limited land availability and supply for primary production. In some parts of the Northeast, Midwest and South of the country, offer for the cultivation land will be reduced.

In general, for the solar energy, it was observed that Brazil had area and highly expressive power for the use of this power, both today and in the coming decades, especially in the North, Northeast and Midwest. In statistical terms, the Mann-Kendal test and Sen's Bend point to a very weak tendency to useful radiation indicator in all regions of Brazil. In addition, it is projected a significant increase in mean air temperature by the end of XXI century across the country that can mean a reduction in power conversion capability, which is sensitive to ambient temperature variations, especially in the Midwest and North of the country.

Finally, it should be noted that this work is based on current technology, thus technology improvement with biotechnology incentives can reverse the results once presented, and these are the recommendations of this study: 1) investment in biotechnology and genetic improvement of plants for better adaptation to climatic contrasts; 2) investment in technology to better use of solar energy for power generation, in addition to the improvement of the materials to be less susceptible to temperature variation and transmission systems, which can reduce energy losses in the distribution process.

ACKNOWLEDGEMENTS

The authors acknowledge the support of the Brazilian Ministry of Science, Technology and Innovation within the "Preparation of the Third National Communication of Brazil to the Framework Convention of the United Nations about Climate Change" project. The authors also thank Rede CLIMA and Sub-Rede Energias Renováveis for the assistance.

REFERENCES

1. Intergovernmental Panel on Climate Change—IPCC: Climate Change (2013) The Physical Science Basis. Contribution of Working Group I to the Fifth Assessment Report of the IPCC. In: Stocker, T.F., Qin, D., Plattner, G.-K., Tignor, M., Allen, S.K., Boschung, J., Nauels, A., Xia, Y., Bex, V. and Midgley, P.M., Eds., Chapter: Summary for Policymakers, Cambridge University Press, Cambridge, United Kingdom and New York, 1535 p.
2. National Electric Energy Agency—ANEEL (2008) Atlas de Energia Elétrica do Brasil. Agência Nacional de Energia Elétrica, Brazil.

3. Gordon, C., Cooper, C., Senior, C.A., et al. (2000) Simulation of SST, Sea Ice Extents and Ocean Heat Transport in a Version of the Hadley Centre Coupled Model without Flux Adjustments. Climate Dynamics, 16, 147-168.
4. Mesinger, F., Chou, S.C., Gomes, J.L., et al. (2012) An Upgraded Version of the Eta Model. Meteorology and Atmospheric Physics, 116, 63-79.
5. Chou, S., Lyra, A., Mourão, C., et al. (2014) Evaluation of the Eta Simulations Nested in Three Global Climate Models. American Journal of Climate Change, 3, 438-454.
6. Chou, S., Lyra, A., Mourão, C., et al. (2014) Assessment of Climate Change over South America under RCP 4.5 and 8.5 Downscaling Scenarios. American Journal of Climate Change, 3, 512-527.
7. Hare, F.K. (1983) Climate and Desertification: A Revised Analysis. World Climate Applications Programme. World Meteorological Organization, Geneva, 149 p.
8. Klein, S.A. (1977) Calculation of Monthly Average Insolation on Titled Surfaces. Solar Energy, 19, 325-329.
9. Sneyers, R. (1975) Sur L'analyse Statistique des Series Dóbservations. Organisation Méteorologique Mondial, Genève, 192 p.
10. Goossens, C. and Berger, A. (1986) Annual and Seasonal Climatic Variations over the Northern Hemisphere and Europe during the Last Century. Annales Geophysicae, 4, 385-400.
11. Sen, P.K. (1968) Estimates of the Regression Coefficient based on Kendall's Tau. Journal of the American Statistical Association, 63, 1379-1389.

5
CHAPTER

SUSTAINABILITY ISSUES AND OPPORTUNITIES IN THE SUGAR AND SUGAR-BIOPRODUCT INDUSTRIES

Gillian Eggleston* and Isabel Lima

USDA-ARS-Southern Regional Research Center, 1100 Robert E. Lee Boulevard, New Orleans, LA 70124, USA

ABSTRACT

Like many other industries, the sugar and sugar-bioproduct industries are facing important sustainability issues and opportunities. The relatively low and fluctuating profit for sugar, surpluses of sugar, world-wide trend to produce alternative, renewable bio-based fuels and chemicals to those derived from petroleum and reduce greenhouse gases, water- and energy-intensive factories and refineries, and increased consumer demands for sustainably manufactured products are putting pressure on the industries to diversify for sustainability. Sugar crops, including sugar and energy cane (*Saccharum officinarum*), sugar and energy beets (*Beta vulgaris*), and sweet sorghum (*Sorghum bicolor* L. Moench), are excellent, renewable biomass feedstocks because of their availability, their being amongst the plants that give the highest yields of carbohydrates per hectare, and high sugar contents. While much research has been focused on conversion technologies for advanced biofuels and bioproducts, attention is now focused on developing sustainable supply chains of sugar feedstocks for the new, flexible biorefineries, with customers wanting maximum feedstock reliability and quality, while minimizing cost. All biomass from sugar crops are potential feedstocks. The cogeneration of bioelectricity from bagasse and leaf residues is being increasingly manufactured in more countries and, due to the high carbon content of bagasse and leaves, can also be converted into value-added products such as biochar. Sugar crops are superior feedstocks for the production of platform chemicals for the manufacture of a range of end-products, e.g., bioplastics, chemicals, and biomaterials. In several countries and regions, green sustainability criteria are now in place and have to be met to count against national biofuel targets. Processes to convert high-fiber sugar crop biomass into biofuel have been developed but there has only been limited commercialization at the large-scale.

Keywords: renewable sugar crops; biomass; advanced biofuels; bioproducts; bagasse; extraneous matter

1. INTRODUCTION

Sucrose (α-D-glucopyranosyl-(1 → 2)-β-D-fructofuranose) is ubiquitously known as common table sugar, and crystalline sucrose is primarily produced industrially from sugarcane (*Saccharum officinarum*) and sugar beet (*Beta vulgaris*) (Figure 1). Like many other food and chemical industries, the sugar industry and related sugar-bioproduct industries are currently facing tough sustainability issues. Sustainability is the balancing of the three, interdependent, development pillars of the environment (ecology), society, and economy (Figure 2). For some industries the core principles for sustainable manufacture are renew, reuse, and recycle, which are applied to every production step and business practice [1]. Sustainable development should also ensure that the needs of the present are met without compromising the ability of future generations to meet their own needs [2].

Although the twentieth century saw enormous growth in chemicals manufacturing which fed the parallel growth in the developed world, it came at a cost. Inefficient processes reliant on fossil fuels leading to unacceptable levels of pollution, hazardous operations resulting in a number of well-publicized disasters, inadequate product testing causing often irrational public concerns over product safety, have all led to an exponential growth in chemicals legislation [3]. Chemical industries, including the sugar and associated industries, are now working towards achieving environmentally acceptable and economically viable manufacturing in a tough legislative framework while meeting the high demands of a growing population. Sustainable production of sugar, biofuels (for example, first generation ethanol in Brazil) and other bioproducts (such as chemicals and structural materials) from sugar crops, will only be realized through a re-assessment of the entire chemical product life-cycle from resources, to manufacturing and production, through to product use and ultimate fate [3,4]. Moreover, several critical changes are required both in mindset and practice that are listed in Table 1.

FIGURE 1. Sugarcane harvested into billets (**top**) and sugar beets being delivered for processing (**bottom**).

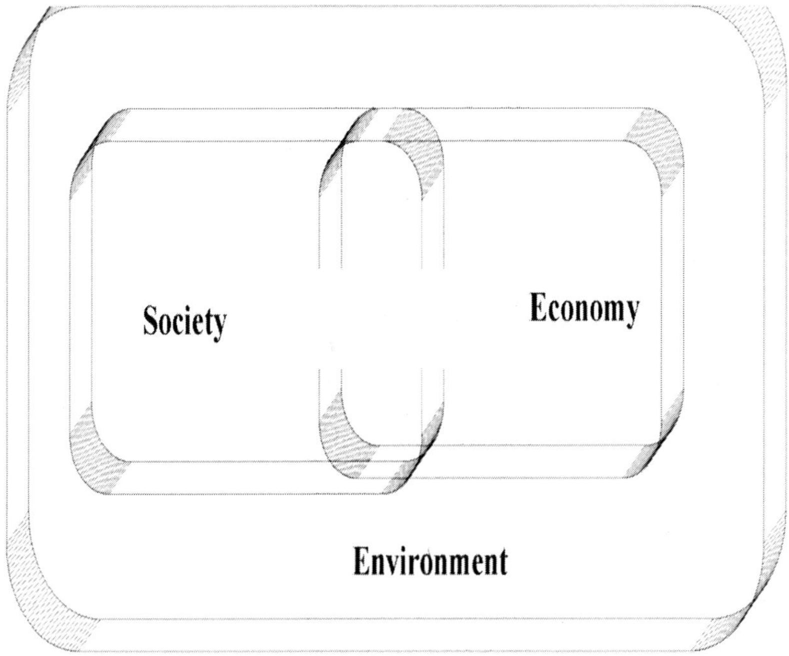

FIGURE 2. Sustainability focuses on the triple bottom line: (i) social responsibility; (ii) economic viability, which are both constrained by (iii) environmental limits that need protection. Sustainable development should recognize the well-being of human systems that is supported by a healthy, natural environment, and which future generations have an equal claim on our planets' resources.

TABLE 1. Unsustainable *versus* sustainable mindsets and practices in the current sugar industry. Adapted from [4].

Key Dimension	Unsustainable	Sustainable
Society/Policy Goals	Economic growth	Growth in well-being
Approach to Nature	Control over nature	Work with nature
Predominant Work Mode	Big is Better	Smart is Better
Focus on Business Activities	Goods	Services, needs
Energy Sources	Fossil fuels	Renewable energy (including biofuels and bioproducts)
Predominant Chemistry	Energy intensive	Low energy
Waste Production	High waste	No waste
Typical Materials	Iron, steel, and cement	Bio-based materials

Continued reliance on fossil fuel energy resources is unsustainable because of depleting world reserves and associated greenhouse gas (GHG), as well as energy security. This explains countries around the world legislating to curb GHG emissions. It also explains the currently vigorous initiatives of developing renewable and potentially carbon neutral, solid, liquid and gaseous biofuels as alternative energy resources as well as biobased alternatives to petroleum-derived chemicals and materials. Biomass, plant-derived organic matter, currently contributes to over 10% of primary energy to meet world annual demand [5] and is expected to grow further [6,7]. In the U.S., the Department of Energy (DOE), via the U.S. Biomass Roadmap, put forward the goal that by 2030 biomass will supply energy approximately equivalent to 30% of current petroleum consumption [8]. The European Commission (EC) has set mandatory targets for an overall share of 20% renewable energy and a 10% share of renewable energy in transport by 2020 [9]. The EC also reached an agreement in 2014 on an indirect land-use change (ILUC) directive [10], to minimize the impact of indirect changes of land use (for biofuel use), while at the same time protecting existing investments in biofuel production in Europe.

Two major challenges that need to be overcome for the creation of renewable biofuels and bioproducts from sugar crops are the need for: (i) net energy gain during feedstock production; and (ii) GHG emissions that are lower than those from fossil fuels [7]. This paper describes current trends, needs, and opportunities in the sugar and sugar-bioproduct industries that are expected to strongly contribute to their sustainability.

2. BACKGROUND INFORMATION: INDUSTRIAL PRODUCTION OF SUGAR AND ASSOCIATED BY-PRODUCTS

2.1. Crystalline Sugar Manufacture

Commercially available sucrose has very high purity (>99.9%) making it one of the purest organic substances produced on an industrial scale. To obtain such a pure product from both sugarcane and sugar beet, complex isolation and purification process units are followed. Industrial sucrose production is essentially a series of separations of non-sucrose compounds (impurities) from sucrose (Figure 3). Sugarcane is grown in tropical and sub-tropical areas of the world and processing often occurs in two stages. Juice is first extracted from sugarcane (sucrose yields range between 10%–15% weight of sugarcane) by tandem milling or diffusion and converted to raw sugar (~97.5%–99.5% pure sucrose; golden yellow/brown crystals) at factories. Secondly, after raw sugar has been transported to a refinery, it is refined using similar unit processes used in raw sugar manufacture, as well as additional decolorization steps such as ion-exchange resins and activated carbon, to the familiar white, refined sugar (>99.9% sucrose) [11].

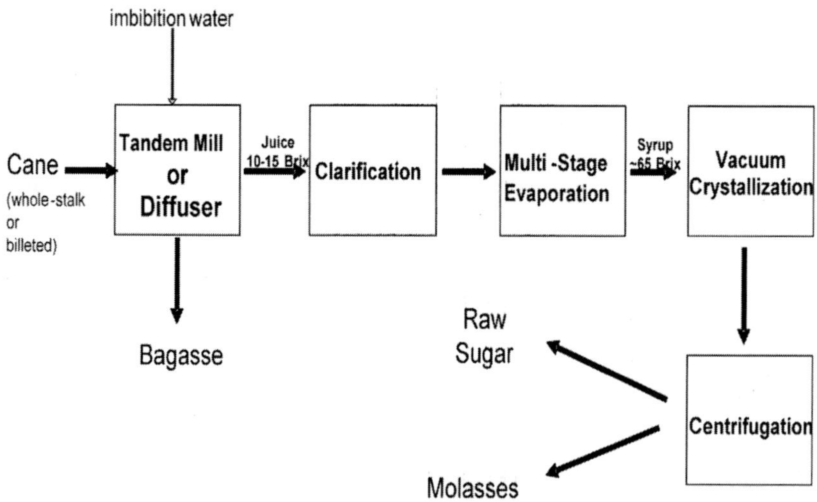

FIGURE 3. Basic scheme of the raw sugar manufacturing process in a sugarcane factory [11]. Brix is % dissolved refractometric solids.

Sugar beets are grown in more temperate areas and processed directly into white sugar (>99.9% sucrose) at nearby factories. Production of refined sugar from sugar beets has some similarities to refined cane sugar production, but dissimilarities exist because sugar beet is a tuberous root and sugarcane a grass. Sugar beets are introduced to the factory, washed, and sliced into "V" shaped cossettes. Sucrose and impurities are extracted from cossettes with hot water in a diffuser. Diffusion juice contains ~12% sucrose and 2% soluble impurities on sugar beet weight, and is heated to ~85 °C before it is clarified. The resulting "thin" juice is then concentrated from ~14 to 60–65 Brix (% dissolved refractometric solids) syrup or "thick juice" across multiple-effect evaporators, then triple-crystallized and centrifuged to produce white, refined sugar (>99.7% purity). As in sugarcane refineries, some sugar beet factories employ additional purification steps. For more detailed information on the industrial production of sucrose from sugarcane and sugar beet, the reader is referred to other comprehensive texts [11,12,13,14,15].

2.2. By-Products of Sugar Manufacture

The major by-products of crystalline sucrose manufacture are sugarcane bagasse, beet pulp, and sugarcane/beet molasses. The major agricultural residue is sugarcane extraneous leafy material. Minor by-products include fly ash, filter cake, lime and calcium carbonate residues. By volume, fibrous bagasse is the most important by-product and is the primary source of fuel for the generation of steam and electricity to operate sugarcane factories (see Section 4.1). Wet and dry beet pulp as well as pressed pulp silage, with or without added molasses, are sources of animal feed. Molasses is a valuable by-product of sugar manufacture and exists in a range of grades: edible molasses, cane and beet molasses, and refinery molasses. It is used as an animal feed additive, in the industrial production of rum and other beverage alcohols, bakers' yeast, citric acid, and other fermentation processes [4].

3. SUSTAINABLE SUPPLY CHAINS OF SUGAR BIOMASS FEEDSTOCKS FOR THE MANUFACTURE OF ADVANCED BIOFUELS AND BIOPRODUCTS

3.1. Sugar Crops as Biomass Feedstocks

A major trend in the U.S. and world-wide is to manufacture advanced biofuels and bioproducts from sugar crops, including sweet sorghum (*Sorghum bicolor* L. Moench), energy cane (*Saccharum officinarum*) and energy beets (*Beta vulgaris*). Sugar crops and their associated by-products and residues make good renewable carbohydrate feedstocks because they are readily available, can be grown in a much larger area of the world than grain crops [4], and are amongst the plants giving the highest yields of carbohydrates per hectare. A unique advantage sugar crops have over grain and cellulosic crops, are that they require less processing as their juice sugars are directly fermentable. Moreover, concerns of conversion efficiency, high production costs, and increased prices for whole corn grain and dried distiller's grain in the USA have provided a real opportunity for alternative feedstock sources from sugar crops [16]. Sugarcane and sugar beet industries, furthermore, have well-established agricultural production systems with a well-developed logistics and processing structure.

At present, approximately 40% of the world's fuel ethanol production is already from sugar crops (mostly the fermentation of either sugarcane juice or molasses in Brazil, Thailand, and India) with the remaining 60% from grain crops [17]. Thus, as of 2014 most fuel ethanol manufactured around the world is first generation ethanol. Also, since 2011, world fuel ethanol production has been steadily rising, with the latest reported figures for current production of 24,570 million gallons in 2014 [18]. The Environmental Protection Agency of the U.S. government designated Brazilian sugarcane ethanol as an advanced biofuel in 2010 due to its estimation that 61% reduction of life cycle GHG emissions, including direct and indirect land-use change emissions [19]. It has been acknowledged, however, that first-generation fuels have technical issues, such as their oxygen contents, that limit their use because they cannot completely replace fossil fuels on their own. For example, 2-carbon ethanol has lower energy per volume than petroleum fuels, and is not fully compatible with existing vehicles and the current fuel-distribution infrastructure [20]. Biofuels with high carbon contents, e.g., 15-carbon alkene β-farnesene that is converted from sugar by engineered yeasts by Amyris in California, offer more sustainable substitution of fossil fuels in the future. Moreover, β-farnesene addresses reduction in GHGs while also delivering improved engine performance [20]. Very recently, Balakrishnan *et al.* [21] went even further and showed that sugars from sugarcane can be converted in >95% yields to a new class of cycloalkane compounds used for aviation fuels and also achieved net life cycle GHG savings of up to 80%.

While much research has been focused on conversion technologies for advanced biofuels and bioproducts, attention is now focused on developing sustainable supply chains of sugar feedstocks for the new, flexible biorefineries [22]. This includes improved feedstock quality and cost-effective approaches for minimizing feedstock sugar losses during storage [22]. Kenney *et al.* [23] similarly reported that while "much progress has been made in improving biomass collection and pre-processing machinery performance and efficiencies, reducing material losses throughout the supply chain, and expanding harvesting and storage operational windows" and "emphasis on feedstock quality is still lacking." The quality of the feedstock supply chain from the field to the processed end-product is impacted by crop genotype variability, production conditions, harvest method, collection and storage practices, season date, and environmental conditions. These relationships and their integration are illustrated in Figure 4 using the examples of sweet sorghum and energy beets, but other sugar crops such as energy cane could be included as well as by-products and residues. Furthermore, as the current pioneer biorefineries move from technology development and deployment to operations, the quality and specifications of the feedstock will become even more important [23]. Overall, customers of sugar feedstocks want maximum feedstock reliability and quality, while minimizing variability and cost.

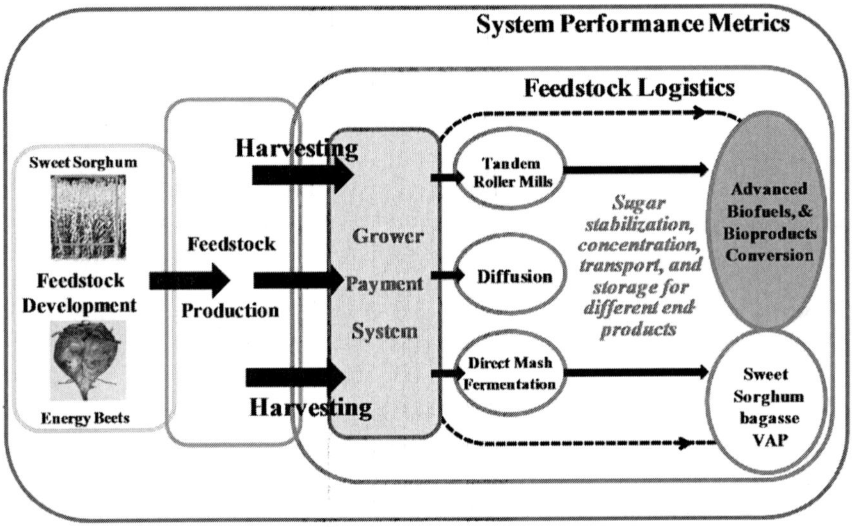

FIGURE 4. The manufacture of advanced biofuels and bio-products supply chain components for sweet sorghum and energy beets. VAP = value added products. This schematic is also applicable to energy canes, bagasse, and extraneous leafy matter.

3.2. Sweet Sorghum as a Biomass Feedstock

Sweet sorghum is a type of sorghum that, like sugarcane, contains juice rich in soluble sugars as well as fiber but, unlike sugarcane, also contains starchy grain that could be used for food, animal feed, or non-food products. Sweet sorghum is an attractive biomass feedstock because it can overcome many of the shortcomings of other sugar crops, due to its efficient C4 photosynthetic pathway, easy cultivation from seed, low fertilizer and water requirements, growth on marginal lands, short growth cycle (90 to 150 days after planting, depending on cultivar and environment) that allows the possibility of multiple crops per season or rotation with other food or non-food crops, wide geographical suitability, and huge breeding potential [24]. It must be noted, however, that due to its low juice purity (sucrose/Brix × 100) of only ~75%, which is markedly lower than for sugarcane (~85%) and sugar beets (~87%) as well as high reducing sugar content, it cannot be used to manufacture crystalline sucrose [25]. Although many current global applications of sweet sorghum are still for small-scale edible syrup and forage production, there has been a recent dramatic increase in interest for large-scale biofuel and bioproduct manufacture [24,26,27]. Several private-sector groups in the U.S. and world-wide have been pursuing development of new domestic industrial sugar feedstocks from sweet sorghum to supply the bioprocessing demand, and new biorefineries are now becoming a reality. For example, Heckemeyer Mill in rural Sikeston, Missouri, recently built the largest, operational sweet sorghum biorefinery in the U.S., which is capable of crushing up to 82 tonnes/h (equivalent to ~49 ha/day) and producing 24,000 gal of syrup (80% solids)/day [28]. Moreover, as sugarcane factories can sit idle for up to 9 months of the year, processing of sweet sorghum to syrup in factories before the sugarcane harvest would allow the use of under-utilized capital equipment. This is currently being explored and tested in numerous sugarcane factories/distilleries in Brazil and by Okeelanta, Florida [29].

4. BAGASSE FROM SUGAR CROPS: LARGE-SCALE COGENERATION OF ELECTRICITY AND OTHER ENERGY PRODUCTS

4.1. Cogeneration of Electricity from Bagasse

Bagasse, the fibrous fraction remaining after juice extraction from sugarcane and sweet sorghum, varies in composition depending on intrinsic and extrinsic factors such as genotype, maturity (lignifications), and environment [30]. Sweet sorghum bagasse is available to a much lesser extent than sugarcane, and its higher amounts of protein can make it more valuable as an animal feed [24]. Most sugarcane processors burn bagasse to cogenerate steam and electricity for running the factory [31]. The recovery of energy from bagasse is also a major reason that sugarcane has a higher net energy ratio (output/input) than many other crops [7]. Furthermore, cogeneration contributes to sustainability as the negative environmental impact of GHGs from traditional thermal power stations are reduced [32]. Presently, some countries' sugar industries, e.g., Brazil, Mauritius, India, Australia, El Salvador, Nicaragua, Guatemala, Columbia, and The Philippines, also operate large-scale cogeneration of electricity and sell the surplus to the local or national grid, and there is great potential for many other countries to follow. Currently, bagasse-based cogeneration is the most profitable sector in the large Brazilian sugar industry, with underperforming sugar and ethanol sectors being bailed out and/or supported recently by cogeneration units [33].

The success of the cogeneration of bagasse depends on the availability of adequate technology as well as a profitable price structure [34]. In recent years, there have been major technological improvements leading to higher efficiency cogeneration of electricity from bagasse, in particular the use of new high-pressure boilers, *i.e.*, up to 82–100 bar (producing superheated steam at 525 °C) [35]. Efficiency gains leading to a surplus of electricity generation for export to the grid have also been accomplished through the retro-fitting of turbo-alternators with high steam pressure and temperature [31], and the optimization of other process parameters, including steam consumption, increasing fiber content of sugarcane through breeding, lower moisture content of bagasse, and reducing the consumption of electricity in the factory tandem mill and power plant [32]. In Brazil, during the 2009/2010 harvesting season the total electricity produced from sugarcane bagasse was 20,031 GWh and this value may rise up to 68,730 GWh over the next 9 years, as long as all factories install 99 bar boilers and 1.04 billion tons of sugarcane is produced [35].

4.2. Second Generation Biofuels from Bagasse

The sustainable trend towards electricity generation from bagasse has been accompanied by progress in developing other large-scale uses for the material, e.g., second-generation ethanol and other biofuels [30] and syngas from Fischer-Tropsch (FT) processes [36,37]. Lignocellulosic ethanol involves the conversion of cellulose and hemicelluloses, which is much more difficult than the conversion of starch and simple sugars. On the other hand, ethanol produced from lignocellulosic feedstocks is seen as a viable option to decrease any perceived competition between the production of foods and bioenergy [30], although energy will always be needed for food production. Other advantages over first generation ethanol are: (i) lignocellulose and cellulose are abundant and less expensive than agricultural food feedstocks; (ii) growth potential is huge; and (iii) some cellulosic crops can be grown in marginal lands that often require less fertilizer and water inputs but grow better on good quality land with optimal imputs.

There are five key-steps for the production of lignocellulosic ethanol from bagasse: (i) milling; (ii) chemical or physical pre-treatment; (iii) enzymatic hydrolysis or saccharification; (iv) fermentation of hexose

(C6) and pentose (C5) sugars; and (v) distillation-dehydration of the ethanol. However, the processing technology for conversion in the most part has only reached very limited commercial scales. Dedini S/A Indústrias de Base, was the first company to build a bagasse-based pilot-scale ethanol facility in Brazil, but commercialization has yet to take place [38]. Gran Bio established the first second generation ethanol facility in Brazil in 2014 in Alagoas. Raizen now has a second generation ethanol facility in Piracicaba, Brazil, that started operating from July 2105. The commercialization of second generation bioethanol depends mostly on economic factors such as values for agricultural feedstocks that have been estimated to range between 50%–80% of the total ethanol's cost [39], government tax incentives for ethanol production, and mandatory ethanol/gas blends [40]. Dantas *et al.* [41] in their assessment of the productivity and cost of different technological routes, determined that, when compared to ethanol production, burning bagasse to generate electricity provided the most benefits from an investor perspective, although this may not be applicable to every country. For more information on this lengthy topic see [41].

4.3. Biochar

Biochars (Figure 5) can be produced via the thermo-chemical conversion of organic feedstocks. Thermo-chemical conversion reactions encompass processes such as: (1) gasification; (2) slow pyrolysis; (3) fast pyrolysis; (4) hydrocarbonization; and (5) combustion, and distinctions among these different options are related to the relative availability of oxygen, residence time, temperature and pressure. Slow pyrolysis (residence time is minutes or hours), in particular, is feedstock-flexible, and this has resulted in the research of a multitude of biomass materials as possible precursors due to their lower cost and availability. Yaman [42] published a review of pyrolysis of various biomass types which includes pyrolysis conditions. Sugarcane bagasse, due to its high carbon content, can serve as an excellent biochar feedstock. Within the thermo-chemical platform, pyrolysis generates biochar as the main product, synthesis gas and bio-oil from the non-condensable fraction of the gas with the split between liquid, char and gas being governed to a certain degree by variation in process conditions. During thermal decomposition of the organic material under limited supply of oxygen and a regime of high temperature (300–700 °C), the material undergoes a series of cleavage reactions, as volatile matter evolves, resulting in a porous high carbon product (Figure 6), mainly composed of aromatic compounds characterized by rings of 6-C atoms linked together. Biochar is part of the black carbon continuum with variable properties due to the net result of production (e.g., feedstock and pyrolysis conditions) and post-production factors (storage or activation). Therefore, biochars are not a single entity but rather span a wide range of black carbon forms [43].

FIGURE 5. Biochar in various forms (pelletized, granular [18 × 40 mesh] and powdered [less than 100 mesh]).

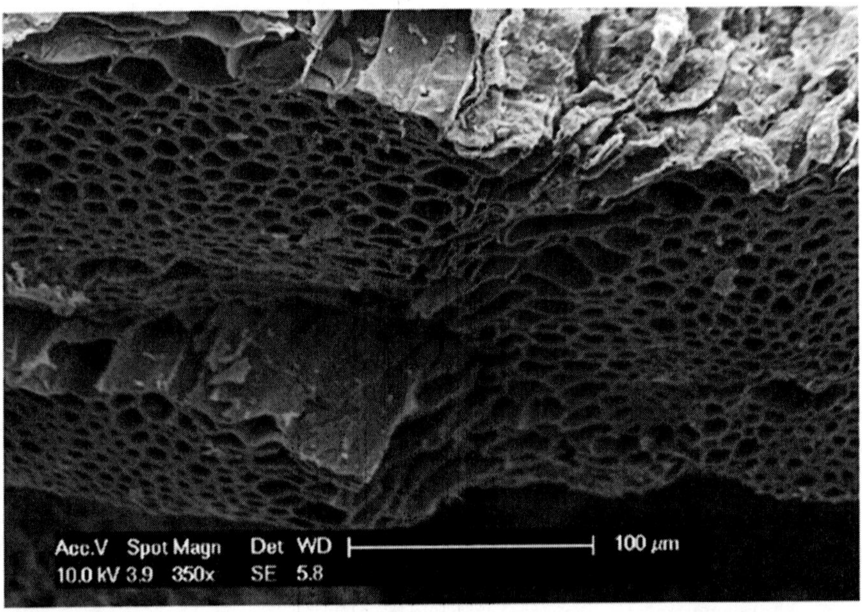

FIGURE 6. Scanning Electron Micrograph of a sugarcane bagasse biochar (magnification of 350×), showing its highly porous nature.

The terminology of biochar is commonly associated to its use in soil applications, and is unlike charcoal, which traditionally was manufactured from coal, and used as a fuel. The first use of the term biochar was around 1998 for the solid residual of biomass pyrolysis [44]. Biochar can be further upgraded or converted into activated carbon through either physical or chemical activation to produce a highly porous material that can be used in adsorption applications.

4.3.1. Biochar as a Soil Amendment

Biochar has been applied to soils virtually from the dawn of civilization, since fire pits were built on soil [43]. The current application of biochar to soil has been modeled after the Amazonian Terra Preta soils, which have higher soil fertility believed to result from intentional additions of biochar from "slash and char" agricultural practices [45,46,47]. Resulting biochars' nutrient contents are variable because they are based on feedstock and production conditions, however, differences have been noted in the chemistries of various biochars despite the similarity in production conditions [48]. Biochar consists mostly of carbon, hydrogen, and oxygen, but also contains the majority of the inorganic micronutrients that were native to the original feedstock. Thus, for biochars made from sugarcane bagasse, the mineral content is transferred to the biochar and concentrated due to burn-off of carbon moieties.

Inorganic material located at the surface of the biochar contributes to surface functional groups and charge that can advantageously enhance microbial relationships in soil as well as contribute to soil's nutrition [49]. Microbial biomass appears to increase in the presence of biochars, however, there is still little insight into the mechanisms by which biochar influences soil microorganisms, fauna, and plant roots [50]. When applied to soil, biochars have also been shown to suppress GHGs such as methane and nitrous oxide from soil [51]. In a recent greenhouse study, Lima *et al.*[52] observed that, by adding 4% *v/v* sugarcane bagasse biochar to soil, the sugar yield of sugarcane improved by 25% over the control. Furthermore, significant increases in sugarcane crop yield were also observed.

4.3.2. Biochar as Fuel

Pyrolysis of biomass including both agricultural and processing residues, as well as forestry products, has been traditionally used to make fuel or charcoal. When considering fuel production, densification technologies such as pelletization and agglomeration can improve the properties of the biomass feedstocks while generating a harder, less abrasive, and higher quality denser fuel. Zandersons *et al.* [53] produced charred fuel briquettes from sugarcane bagasse and determined that volatile matter given off during pyrolysis was sufficient to provide drying heat as well as sustain the pyrolysis process. Molasses is commonly used as a binder and conveniently available at the factory, making co-location of biochar manufacturing facilities in proximity to the factory even more efficient. Because sugarcane bagasse is considered a softer biomass material when compared to other biomass materials such as wood, Pendyal *et al.* [54] used sugarcane and corn syrup molasses as binders to create briquettes before pyrolysis. The suitability of bagasse biochar as a fuel, either by itself or blended with other by-products (e.g., leafy material) depends, among other parameters, on its heating value. A high carbon content, reported as 47% on a moisture and ash free basis, is also desirable and directly affects the fuel value [55], while ash content negatively impacts fuel value. Several empirical correlations have been established to calculate heating value from either proximate analysis (volatile matter, fixed carbon and ash content) [56] or from the elemental composition (e.g., C, H, N, O) [57]. Based on several available reports in literature, dry bagasse contains 82%–87% volatile matter, 1%–7% ash, and 10%–17% fixed carbon [58,59,60,61].

4.3.3. Biochar as an Adsorbent

Extensive literature exists on the use of biochars and their activated counterparts in remediation applications. Biochars produced from biomass such as sugarcane bagasse have low surface area when compared to commercial activated carbons, but their surface functionality can make them an attractive alternative particularly in adsorption of charged species. Additionally, they have a relatively lower cost of production, due to higher yield and more simple manufacture. When studying various agricultural by-products blended with various types of molasses, Ahmedna *et al.*[62] found that sugarcane bagasse was a better choice than rice straw or rice hulls as a precursor for activated carbons with desirable sugar decolorization properties, especially when combined with corn syrup. In another study, Ahmedna *et al.* [63] found that bagasse activated carbons performed as well as commercial carbons commonly used in sugarcane refineries, despite the fact of having half as much surface area. Because color removal both at the sugarcane factory and particularly refinery is such an integral part of the process, it is logical from a sustainability viewpoint to utilize biochars from sugarcane bagasse and/or trash as alternate color removal adsorbents. Feasibility studies for producing activated carbons from sugarcane bagasse and molasses revealed the cost to be \$3.12/kg of carbon for a 2000 kg daily output [64,65]. In heavy metal ion adsorption applications, a sugarcane bagasse activated carbon outperformed a commercial carbon in the adsorption of nickel, copper, zinc, and lead, despite having much lower surface area [66]. This was most likely because of the presence of surface oxides such as carbonyls, lactones, phenols, and carbonyls giving biochars a negative surface charge [63]. Binder choice can also play a significant role in the biochar properties although properties of the base material will ultimately determine the properties of the biochar and the activated carbon [54].

4.3.4. Biochar for Carbon Sequestration

The process of biochar production has the potential of being carbon negative and, therefore, a carbon sinking technology by which a large percentage of the carbon in the biochars is in the fixed form. As sugarcane grows, it takes in carbon dioxide and converts it into plant building blocks such as cellulose and lignin, which will be again turned back into carbon dioxide through combustion. This process can be altered if carbon in sugarcane bagasse is instead converted into fixed carbon form as happens with pyrolysis. Pyrolysis temperature and residence time are directly proportional to the percent of fixed carbon generated. To effectively reverse climate change it may be necessary to return some of the atmosphere's carbon dioxide to the soil [67]. One way of achieving this, is by pyrolyzing biomass such as sugarcane bagasse to produce gas or oil for energy and using the remaining chars as soil amendments. Benefits of biochars use as climate change mitigation tool was recognized and reported to the U.S. Congress [68].

4.4. Storage of Large Piles of Bagasse

Challenges still exist with respect to the prolonged storage of very large quantities (can be >900,000 tonnes) of bagasse (~50% moisture content), particularly for off-season electricity generation. These include microbial decomposition and loss of fuel value, self-heating leading to spontaneous combustion, bagasse handling, and a variety of health issues, and environmental impacts [37]. An important health issue is bagassosis—A respiratory disease resulting from exposure to fungal spores from moldy bagasse dust. Environmental impacts include: (i) water pollution due to run-off from bagasse piles; (ii) bagasse dust that is a major problem in windy areas and at bagasse transfer points; (iii) noise and lights that have caused problems at storage sites close to residential areas; and (iv) off-odors.

Studies have been undertaken in various countries to overcome storage challenges, and in the case of Australia, guidelines for storage have even been established with cooperation of the Environment Protection Agency [37]. Options to solve bagasse storage problems have been reported and include the patented "Bagatex-20" process developed in Brazil [69], which depends on a biochemical catalyst to accelerate and control fermentation in the piles. Other options are forced drying before storage [70], special stacking, covering of piles [37], and densification processes. The latter includes pelletizing, which is known to significantly reduce final volume with bulk densities in the range of 1030 to 1260 kg/m^3, depending on pellet diameter [55]. Furthermore, pelletizing sugarcane bagasse is a way of improving fuel handling, transportation, conversion and also allowing for storage for off-season utilization [55]. Recent initial results from a sugarcane bagasse storage study in the U.S. [71], indicated that dramatic gains in fuel value can be achieved just by covering stored bagasse piles with tarpaulin (Table 2). Covering the bagasse also reduced the rate of deterioration [71].

Table 2. Proximate Analysis and estimated fuel value of sugarcane bagasse samples as a function of storage conditions across two factories *.

Storage Conditions	Sample	Moisture %	Ash Content %	Fixed Carbon %	VOC %	Fuel Value, KJ/kg HHV [59]	Moisture LHV [72]
Covered	Fresh	38.5	6.0	15.7	78.3	17,712	10,890
	3 months	49.9	8.3	13.6	77.7	16,980	8504
	4 months	43.1	9.0	13.9	77.1	16,857	9590
	5 months	33.6	7.6	15.3	77.1	17,371	11,535
	6 months	56.9	9.9	14.0	76.1	16,738	7213
Uncovered	Fresh	-	-	-	-	-	-
	3 months	81.0	15.7	8.9	75.5	14,779	2805
	4 months	73.7	10.9	11.2	77.9	16,021	4203
	5 months	73.1	17.5	8.8	73.7	14,472	3889
	6 months	68.1	15.2	11.1	73.8	15,289	4875

* Data presented at the 45th Annual Joint ASSCT Meeting in New Orleans, LA 2015 [71].

5. NEW SUSTAINABLE USES FOR SUGAR CROP EXTRANEOUS MATTER (LEAVES AND TOPS)—AGRICULTURAL BIOMASS RESIDUE

5.1. New Uses for Sugarcane Extraneous Matter

Although sugarcane extraneous matter (E.M. also known as trash or leaves and tops) is an agricultural biomass residue that is widely available in many countries, but is still under-used or improperly utilized. There is, however, a growing reality that E.M. represents a rich source of renewable biomass. Sweet sorghum E.M. is also available but to a much lesser extent than for sugarcane, and the higher amounts of protein in sweet sorghum can make it more valuable as an animal feed [24]. For sugarcane, the continuing global reduction in pre-harvest burning of cane leaves as well as the increased use of green mechanical harvesting have considerably increased the availability of E.M. that can be collected either in the field or at the factory [35]. It should be noted, however, that some sugarcane E.M. must be left on the fields to protect the soil and recycle nutrients [73,74], nevertheless there is still plenty left to serve as a biomass feedstock.

Like other sustainable biomass sources, E.M. can be converted through a wide range of biochemical or thermal platform technologies into a multitude of chemicals, bioenergy, and biomaterials including cellulosic ethanol, electricity, and biochar [75]. E.M. also represents a real opportunity for rural communities to access local energy supplies and bring economic opportunities to many developing countries that grow sugarcane [5]. Sugarcane E.M. has already contributed to greater cogeneration of electricity [31,35]. In 2007 a study in Brazil showed that the utilization of sugarcane E.M. with bagasse can double the MWh production of electricity compared to bagasse alone [76]. E.M. can also be combined with bagasse to produce biochar, either with or without pelletizing, and the presence of certain minerals in the E.M. such as calcium, potassium, phosphorous, can be beneficial when using the biochar as soil amendment in sugarcane.

Considerable amounts of E.M. are still delivered to factories and processed to the detriment of the quantity and quality of raw sugar produced [77]. Thus, the development of large-scale industries utilizing E.M. as a biomass feedstock could additionally improve the sustainability and profitability of the sugar industry by indirectly improving the quality of raw sugar. The use of sugarcane E.M. as a biomass feedstock is greatly dependent on the amount of dry mass available after collection. It is known that genotype and environmental variation occurs for sugarcane E.M. Sugarcane in the U.S., has ~34% total dry E.M. biomass [78] compared to 41% for some sugarcane varieties in South Africa [77]. Dried leaf residue contains approximately 36% cellulose, 21% hemicelluloses, and 16% lignin [79] and is similar to the composition of bagasse. The quantity and quality of leaves can also vary across the harvesting season in different countries. Donaldson *et al.*[80] in South Africa and Eggleston *et al.* [81] in Louisiana, U.S., both reported that the amounts of green leaves varied with season date while dried, brown leaves changed little with season.

5.2. Biochar from Extraneous Matter

Along with sugarcane bagasse, E.M. is another feasible feedstock for the manufacture of biochar. Varying amounts of soil, leaves, *etc.*, present in E.M. will, however, produce heterogenous biochars of equally varying physico-chemical properties, which could present a challenge to their usability. Because E.M. is mainly composed of leaves, it is significantly less heavy than sugarcane bagasse, with approximately 4 to 5 times lower densities. This can become an issue during the field application of biochars due to wind losses, as well as during storage. Densification technologies can potentially address these challenges, although the moisture content would have to be reduced prior to pelletization to <20%. As with sugarcane bagasse, sugarcane molasses can be used as a binder in biochar manufacture from E.M. [62,63,66].

5.3. Collection and Storage of Sugarcane Extraneous Matter

If E.M. is to work as a reliable, sustainable, and economical biomass feedstock to the new biorefineries, then collection of excess E.M. in the field needs to be optimized, preferably after some solar drying to create greater dry mass [82]. Collection could include baling to increase the material's bulk density for transportation to the biorefinery. As an alternative, E.M. could be separated from the stalks at the biorefinery site if economical and effective separation technologies are in place such as dry cleaning before the sugarcane is shredded [83]. Alta Mogiana sugarcane factory in Sao Paulo State, Brazil has an E.M. dry cleaning system and the separated E.M. is burnt in the boilers. In 2014, NFR BioEnergy LLC, initiated an E.M. dry cleaning system of their own design in the USA, which is co-located with the Cora Texas sugarcane factory in White Castle, Louisiana [84]. NFR BioEnergy already has the technology to produce biochars from sugarcane bagasse and plan to also use separated E.M. as a feedstock. Their dry cleaning system incorporates a large drying tumbler that effectively removes E.M. from the cane. The collected E.M, together with excess bagasse, will be fed into a torrefaction or pyrolysis unit for the manufacture of biochars, although currently the unit is only pilot-plant size. In their process, synthesis gas will also be recovered and used to supply energy to the torrefier/pyrolysis unit. The objective is to create a new source of sustainable energy from using the pelleted biochar as fuel, but to also remove E.M. and allow for cleaner sugarcane to be processed at the factory. The removal of leaves and soil before it enters the factory will allow more sugar to be produced at a higher quality than was previously possible [85]. Energy generated from syngas released during torrefaction can be used to make the whole system "energy negative" and to help run additional equipment such as a pellet mill upstream from the torrefier [86]. However, questions still remain on how efficiently E.M. separation technologies perform while not removing valuable sucrose in stalks [74,87]. Moreover, in some cases separation at the biorefinery could also create excessively large piles that would have to be stored or rapidly utilized [74].

Just as for bagasse, if E.M. is to be a stable and sustainable biomass feedstock it will most likely be stored in piles at the biorefinery or near the harvest site. Unlike bagasse, however, very little is known on how piles of E.M. store and for how long. Eggleston *et al.* [88] reported on the deterioration of brown, dry (BL) and green leaves (GL), stripped from sugarcane whole-stalks, stored under simulated wet and dry conditions. The worst deterioration for both BL and GL, generally, occurred in the watered samples and when the humidity was highest. On deterioration, more soluble impurities were extracted from GL than BL. Only prolonged deterioration of BL caused a reduction of fiber biomass, and the fiber content of GL usually increased on deterioration because of loss of moisture. Because its shredded state created more surface area to absorb water, moisture was highest in deteriorated BL; this also allowed for more *Leuconostoc mesenteroides*bacteria to grow and form dextran and mannitol [88].

6. DEDICATED ENERGY FEEDSTOCKS FROM SUGAR CROPS

The manufacture of second generation biofuels is expected to utilize a much more diverse set of feedstocks compared to first generation biofuels [89], and dedicated energy crops represent one option. Dedicated energy crops offer high output per hectare with low inputs. They are also expected to grow on land less suitable for food production. Often, the new energy crops yield well under various stress conditions and have even been developed with advantageous processing characteristics [90]. Energy cane, beet and sorghum crops can be converted to second generation cellulosic fuel ethanol as well as power and bioelectricity. Processes to convert high fiber energy canes and beets into fuel ethanol are still under investigation as discussed in Section 4.2. The challenge is to develop energy crops with a suite of desirable physical and chemical traits while increasing biomass yields by a factor of two or more [91].

6.1. Energy Canes

Energy cane varieties are high-fiber and biomass clones of conventional sugarcane [92], and often have extended ratooning abilities. The goals of achieving energy cane with attributes for a biomass-based economy are readily achievable through existing sugarcane breeding programs [7]. In sugarcane breeding, more rapid genetic gain can occur for total biomass than sugar yield because growth does not have to be intentionally restricted during the life cycle of the crop, and a wider array of germplasm of potential value is available to the breeder once stringent standards for sucrose and fiber levels are relaxed. Breeding strategies for energy cane are classified as Type I and Type II. Type I are energy cane varieties close to conventional sugarcane but have lower sucrose content and thus higher fiber contents. Type II are energy cane varieties that have marginal sugar content and high fiber at such high levels (typically >16%) that raw sugar manufacturers consider unacceptable for processing [7]. Furthermore, the Type II energy canes would be utilized as biomass feedstock for the production of electricity and cellulosic biofuels [92]. For more information about Type I and Type II energy canes the reader is referred to Botha and Moore [7].

A few energy cane varieties have already been developed and released, for example in Louisiana, U.S. [93]. Three Type I high fiber sugarcane varieties (L 79-1002, HoCP 91-552, and Ho 00-961) released for commercial planting in 2007 produce dry biomass yields in excess of 25 tonnes/ha [93]. As marginal land to grow energy canes in Louisiana are mostly north and, therefore, colder during the winter, a major emphasis of the breeding program is to breed for cold tolerance.

6.2. Energy Beets

Many biofuel/bioproduct sugar feedstock producers, including U.S. large-scale sweet sorghum producers, are considering energy beets as a co-rotation crop to allow the year round production of sugar feedstocks. Energy beet is a non-edible, water efficient, selected hybrid of commercial sugar beet that has not been bred for sugar but for dry matter. The creation of higher biomass yields for energy beets have been made possible by using fodder beet germplasm as a parent in hybrids with sugar beet [94,95]. Biomass yield potential is dependent upon interception of solar radiation which gives beets grown in areas with long growing seasons a decided advantage. Energy beets are typically grown as winter crops (winter beets in the U.S. have a longer growing season and, therefore, a much higher yield potential), can grow on marginal lands, and also have higher fiber, glucose and fructose concentrations [96]. However, the glucose and fructose concentrations are still lower than in sweet sorghum, and the sucrose considerably higher. Overall, energy beets hybrids are comparatively new compared to conventional sugar beets, and very little processing research has been reported.

7. SUGAR AND SUGAR-BIOPRODUCT INDUSTRIES: PLATFORM CHEMICALS FROM SUGAR FEEDSTOCKS AND OTHER VALUE ADDED PRODUCTS FROM SUCROSE

7.1. Platform Chemicals from Sugar Feedstocks

Several years ago, Novozymes CEO Steen Riisgaard said "in a few years sugar will be the new oil" as sugar is a superb feedstock for the production of platform chemicals for the manufacture of a range of end-products, e.g., bioplastics, industrial solvents, and chemicals [97]. This statement still holds true today and structural bio-based materials, such as artificial spider silk "Spiber™ (Stockholm, Sweden)" which is stronger than steel and lighter than carbon fiber [98], can now be added to the portfolio of possible end-products. Furthermore, Koch *et al.* [99] recently stated that "a promising option for intermediates of sugar production, including syrup, is to ferment them and this is less expensive than pure sucrose". Scientists are also currently taking a closer look at natural sweeteners such as honey, maple syrup, molasses, and sweet sorghum syrup, as sugar replacements. This is because they contain, beyond sucrose, glucose, and fructose, other classes of bioactive compounds including complex carbohydrates, amino acids, and polyphenols that might impart health benefits [20].

Efforts in "green chemistry" have been ramped up to transform renewable crop biomass, e.g., from sugar crops, into the basic chemical ingredients that go into many everyday products [99,100]. Thanks to numerous years of work on engineered microbes and new catalysts, the reach of biobased chemicals into consumer items is expanding [101]. A host of biobased intermediates are at or near commercialization, and include raw materials for common polymers such as polyester, spandex, synthetic rubber, and nylon [101]. Such products are being manufactured by both start-up firms and industrial giants, with announcements of progress gaining in frequency and substance [101]. A large-scale shift to bio-based polymers, however, will depend on the availability and reliability of large quantities of sugar feedstock at competitive prices. Moreover, as Bomgardner [101] stated that commercializing a biobased polymer "requires sustained, parallel progress on several fronts including technology development, end-product verification, market demand, and robust business acumen." In 2004, DOE identified a set of biomass-derived compounds best suited to replace petroleum-derived chemicals [102], and Table 3 lists the current and projected commercialization of these products [101]. Only succinic acid, sorbitol, and xylitol productions are established on a large, commercial scale (Table 3). For example, DSM and Myriant are now manufacturing succinic acid in Italy [103] and the USA, respectively. It must be noted that some of the new, start-up companies that started out aiming for commodity biofuels and bioproducts have ended up manufacturing specialty compounds instead [104]. This is because selling in expensive markets helps pay the company bills when the start-up scale is small and its products are still costly to manufacture. Moreover, this strategy can allow companies to work their way into commodities [104].

Although there is no current, effective one-step method or multi-step methods for converting raw lignocellulose to finished products, progress is being made. Furthermore, increasing investments in the sugar-ethanol industry could facilitate the construction of the physical infrastructure, and associated technologies that could also be used for the production of bioproducts [97]. Biotechnology processes are particularly suited for the transformation of natural feedstock from sugar crops into the necessary sugars and building blocks of secondary bioproducts, and bioethanol itself can also be used as a platform chemical [97].

7.2. Value Added Products from Sucrose

For the sugar industry, value-added products from sucrose can increase the demand, value, and consumption of sucrose, as well as improve the industry's competitiveness. However, only a small percentage of the sugar produced in the world is used in non-food applications [105], which is unfortunate as much research effort and funds have been expanded on the identification and development of value-added products from sucrose. Part of the reason for such little impact of this research is that the scientists inventing the products have not fully considered the market, and do not have the business acumen to sell such products to industry [11]. More involvement by industry, particularly at the conception phase, would help to gain more impact [11]. Another explanation is that, historically, fossil-based oil has been a less expensive feedstock than sugars.

Table 3. Current and Projected Commercialization of Biobased Products from Sugar Feedstocks that were Identified in 2004 by the U.S. Department of Energy (DOE) and the most likely compounds to replace petroleum-derived chemicals. Adapted from Bomgardner [101].

Biobased Compound	Number of Carbon Atoms	Method of Manufacture	Key Uses and End-Products	Commercialization in 2014	Projected Commercialization in 2024
Succinic acid	4	Bacterial fermentation of glucose, chemical oxidation of 1,4-butanediol	Solvents, polyesters, poly-urethanes, nylon, food and beverage acid control, surfactants, adhesives, fabrics, inks, paints	Yes	Yes
2,5-Furandicarboxylic acid	6	Chemical dehydration of glucose, oxidation of 5-HMF	Polymers such as nylon, plastic bottles/containers, carpet fiber	Soon	Yes
3-Hydroxypropionic acid	3	Bacterial fermentation of glucose	Polymers, carpet fiber, paints and adhesives, superabsorbent, contact lenses	No	Yes
Sorbitol	6	Hydrogenation of glucose from corn syrup, bacterial fermentation under development	Sweeteners, fuel ingredients, anti-freeze, water treatment	Yes	Yes
Xylitol	5	Hydrogenation of xylose, extraction from lingo-celluloses, bacterial fermentation under development	Sweeteners, cough drops and medicines, anti-freeze, new polyesters	Yes	Yes
Levulinic acid	5	Acid-catalyzed dehydration of sugars	Fuel ingredients, solvents, plastic bottles, polyesters, polyamides, pharmaceuticals, herbicides	No	Maybe
Itaconic acid	5	Fungal fermentation of glucose	Styrene-butadiene copolymers, rubber, plastics, paper	No	Maybe
3-Hydroxybutyrolactone	4	Multi-step chemical synthesis from starch	Solvents, synthetic intermediates for pharmaceuticals, new polymers	No	Maybe
Glutamic acid	5	Bacterial fermentation of glucose	Polyesters, nylon analogs, flavor enhancers, fabrics, plastics	No	Maybe

Sucrose is a good source for many value-added products because of its chemical and enzymatic reactivity. The basis for the reactivity of sucrose is the eight hydroxyl groups present on the molecule. Generally, the three primary hydroxyls have greater reactivity but they often prove a hindrance as they are difficult to react exclusively [106]. The synthesis of an enormous number of sucrose derivatives is possible; substitution with just one group type could theoretically give two hundred and fifty five different compounds. Moreover, the alcohol group can be derivatized to form esters, ethers, and substitution derivatives [106]. Sucrose can be readily degraded by acids, oxidizing agents, alkalis, and catalytic hydrogen to compounds of lower molecular weight. Sucrose is also an exceptional molecule for enzymatic synthesis reactions [105,107] to form, for example, oligosaccharides and polysaccharides [108].

8. VHP (VERY HIGH POL) AND VLC (VERY LOW COLOR) SUGAR PRODUCTION—A SUSTAINABLE TREND

In recent times there has been a world-wide trend to manufacture VHP and VLC raw sugars for supply to refineries, *i.e.*, a trend of vertical integration from the field to the white sugar output. Furthermore, a concomitant trend exists to build refineries of the VHP/VLC cane raw sugar close to the consumption areas to satisfy the needs of the food industry. There is also a growing demand for exports of VHP and VVHP (very, very high pol) raw sugars, particularly from Brazil, mainly for overseas markets. Some refineries also want lower ash concentrations in the VHP/VLC sugar because: (a) some of the refined sugar will be manufactured into liquid sugar, which requires low ash; and (b) lower ash is needed for short, medium, and long term refinery strategies [4]. The supply of higher quality raw sugars is expected to create additional efficiencies at the new refineries, particularly at the early, energy-intensive affination stage. Some refineries such as Al Khalij Sugar in Dubai and Dangoto in Lagos are even eliminating the affination stage altogether. The higher quality raw sugars will also allow factory processors to gain premiums from the new refineries. Furthermore, manufacture of higher quality raw sugars at the factory where the energy source is renewable bagasse, will save fossil energy utilization by the refiners.

9. SUSTAINABILITY METRICS

Traditionally, chemical/food process and product development has focused on the assessment of economic criteria, but additional criteria for sustainability have become increasingly important and integrated into decision making processes [109]. Assessment tools, standards, and enhanced metrics to measure "green, greener, or greenest" have been and are continuing to be developed [110]. Many countries and regions have introduced policies or adopted standards to promote sustainable manufacture of biofuels and bioproducts and use, most prominently in the U.S. and European Union. Ecological or environmental sustainability, one of the three pillars of sustainability (Figure 2) can be examined using Life Cycle Assessment (LCA) [110]. LCA is an internationally recognized methodology for evaluating environmental performance of a product, process, or pathway along its partial or whole life cycle, which can be applied to new processes for converting sugar biomass [7]. Late in 2014, the World Business Council for Sustainable Development (WBCSD) in Geneva, Switzerland, published a guide titled "Life Cycle Metrics for Chemical Products" to help the chemical industry and associated stakeholders compare, on a common sense basis, the environmental footprint of chemical products [111]. This should allow chemical sector companies, including those using sugar feedstocks, to communicate with a common language to companies downstream, and help scaling up solutions [111]. Current operating chemical companies such as BASF, Amcor, and Dow Chemical are benefiting from using sustainability metrics to meet the challenge of improving processes and products, and have already provided insights on opportunities and progress [112].

Rein [113] specifically reported on measuring and monitoring sustainability in the sugar and sugar-ethanol industries. The Better Sugarcane Initiative (BSI) is a collaboration of sugar retailers, investors, traders, producers, and NGOs who are committed to sustainable sugar production by establishing principles and criteria that can be applied to sugarcane [113]. In the sugarcane industry, Brazil has been the most active in embracing and reporting sustainability performance, mostly because they export to other countries [113], although broader sustainability reporting relating to social and other environmental issues have been less of a focus in Brazil. For the future sustainability of the sugar and sugar-bioproduct industries, there is also a need for new analytical methods in biofuels and bioproducts manufacture as well as for grower payment systems with the new feedstock crops [4].

10. OVERALL OUTLOOK

Biomass from sugar crops is expected to become a significant part of the world-wide shift from a fossil-fuel based economy to one that is biobased [7]. This shift, however, is still mostly in the early pioneering stage. Key criteria in the establishment of commercial biobased plants are cost of the facility, availability of reliable quality feedstocks at low cost, financing, and the policy environment [103]. Concomitantly, in many areas of the world, there is a progressive diversification of the sugar industry into "sugar and sugar-bioproduct industries" that are deeply involved in the maximization of the sugar crop biomass. This includes Europe where many beet factories have diversified to produce ethanol as well as sugar from sugar beets. An excellent example of a sugar company diversifying into biobased chemicals while obtaining social, environmental, and financial sustainability gains, is Godavari in India [114]. Starting with sugarcane, Godavari manufactures refined sugar, ethanol, and chemicals including ethyl acetate, crotonaldehyde, 1,3-butanediol, and even flavor and fragrance ingredients [114]. From bagasse, electricity is cogenerated and used to power the sugar and chemicals factories and plants. Godavari's biobased chemical are produced so efficiently that they are able to compete pricewise against the same chemical produced from fossil sources [114]. It is expected that more "sugar" and "sugar-bioproduct" companies like Godavari, will become more and more eager to become greener [110] as they realize that they can reduce pollution and increase profits simultaneously [115]. Moreover, such companies will want to be able to select greener starting materials and use cleaner chemical processes to make environmentally preferred products [110], and to appease consumers who are becoming increasingly aware of sustainability issues [116]. Consumers are, in fact, holding companies to higher standards and asking them to demonstrate the wider impact of their operations [116].

Overall, the sustainability of the sugar and sugar-bioproduct industries should be viewed as a continuous improvement journey [1], and behavior change and education will also be linchpins in effective sustainability programs.

ACKNOWLEDGMENTS

Mention of trade names or commercial products in this article is solely for the purpose of providing specific information and does not imply recommendation or endorsement by the U.S. Department of Agriculture. USDA is an equal opportunity provider and employer.

AUTHOR CONTRIBUTIONS

Gillian Eggleston led the research on sugar processing, the effect of sugarcane extraneous mass on processing and how it deteriorates on storage. She also conducted the review of literature on sugar processing, use of sugar crops as a biomass source. Isabel Lima led and conducted the research on biochar production from bagasse and extraneous matter, and also conducted the review of literature on these topics.

REFERENCES

1. Higgins, K.T. Tastes great, less carbon. *Food Eng.* **2009**, *9*, 51–58.
2. Brundtland, G.H. *Brundtland Report*; Oxford University Press: Oxford, UK, 1987.
3. Abraham, M.A.; Clark, J.; Winterton, N. Introduction. In *Green Chemistry Metrics*; Lapkin, A., Constable, D., Eds.; Wiley & Sons: Oxford, UK, 2008; pp. 1–26.
4. Eggleston, G. Future sustainability of the sugar and sugar-ethanol industries. In*Sustainability of the Sugar and Sugar-Ethanol Industries*; ACS Symposium Series 1058; Eggleston, G., Ed.; Oxford University Press: Oxford, UK, 2010; pp. 1–19.
5. Chen, P.; Wan, Y.; Wang, X.; Cheng, Y.; Liu, Y.; Lin, X.; Ruan, R. Bioenergy industry status and prospects. In *Industrial Crops and Uses*; Singh, B.P., Ed.; CABI: Wallingford, UK, 2010; pp. 21–34.
6. Jiang, D.; Zhuang, D.; Huang, Y. Crop residues as an energy feedstock. Availability and sustainability. In *Sustainable Bioenergy Production*; Wang, L., Ed.; CRC Press: Boca Raton, FL, USA, 2014; pp. 223–235.
7. Botha, F.C.; Moore, P.H. Biomass and bionergy. In *Sugarcane: Physiology, Biochemistry, and Functional Biology*; Botha, F.C., Moore, P.H., Eds.; Wiley & Sons: Hoboken, NJ, USA, 2014; pp. 521–540.
8. Han, F.X.; King, R.L.; Lindner, J.S.; Yu, T.Y.; Durbha, S.S.; Younan, N.H.; Monts, D.L.; Su, Y.; Luthe, J.C.; Plodinec, M.J. Nutrient fertilizer requirements for sustainable biomass supply to meet US bioenergy goal. *Biomass Bioenergy* **2011**, *35*, 253–262.

9. Van Dam, J.; Junginger, M. Striving to further harmonization of sustainability criteria for bioenergy in Europe: Recommendations from a stakeholder questionnaire. *Energy Policy***2011**, *39*, 4051–4066.
10. Ottewell, S. Europe targets biofuel sustainability. Chemical Processing, 20 November 2014, 58.
11. Eggleston, G. Sucrose and related oligosaccharides. In *Glycoscience*; Fraser-Reid, B., Tatsuta, K., Thiem, J., Eds.; Springer-Verlag: Berlin, Germany, 2008; pp. 1163–1182.
12. Van der Poel, P.W.; Schiweck, H.; Schwartz, T. *Sugar Technology. Beet and Cane Sugar Manufacture*; Verlag Dr. Albert Bartens KG: Berlin, Germany, 1998.
13. Chen, J.C.P.; Chou, C.C. *Cane Sugar Handbook*, 12th ed.; John Wiley and Sons, Inc.: New York, NY, USA, 1993.
14. Hugot, E. *Handbook of Cane Sugar Engineering*, 2nd ed.; Elsevier: Amsterdam, The Netherlands, 1972.
15. McGinnis, R.A. *Beet-Sugar Technology*, 3rd ed.; Beet Sugar Development Foundation: Fort Collins, CO, USA, 1982.
16. Ragagopal, D.S.; Sexton, G.; Hochman, D.; Roland-Holst, D.; Zilberman, D. Model estimates food-versus-biofuel trade-off. *Calif. Agric.* **2009**, *63*, 199–201.
17. RFA: Renewable Fuels Association. World Fuel Ethanol Production. Available online: http://ethanolrfa.org/pages/World-Fuel-Ethanol-Production (accessed on 24 August 2015).
18. F.O. Licht World Ethanol and Biofuel Report. Available online: https://store.agra-net.com/(accessed on 24 August 2015).
19. U.S. Environmental Protection Agency. Greenhouse Gas Reduction Thresholds. 2010. Available online: http://www.epa.gov/OMS/renewablefuels/420f10007.htm#7 (accessed on 28 June 2015).
20. Ritter, S.K. Green chemistry awards. C&E News, 20 October 2014, 32–34.
21. Blakarishnan, M.; Sacia, E.R.; Sreekumar, S.; Gunbas, G.; Gokhale, A.A.; Scown, C.D.; Dean Toste, F.; Bell, A.T. Novel pathways for fuels and lubricants from biomass optimized using life-cycle greenhouse gas assessment. *Proc. Natl. Acad. Sci. USA* **2015**, *112*, 7645–7649.
22. Koninckx, J. Biorefinery beckons: Iowa plant will produce ethanol from corn stover. Chemical Processing, 3 January 2013, 38–39.

23. Kenney, K.L.; Smith, W.A.; Gresham, G.L.; Westover, T.L. Understanding biomass feedstock variability. *Biofuels* **2013**, *4*, 111–127.
24. Eggleston, G.; Cole, M.; Andrzejewski, B. New commercially viable processing technologies for the production of sugar feedstocks from sweet sorghum (*Sorghum bicolor* L. Moench) for biofuel and bioproducts manufacture. *Sugar Tech.* **2013**, *15*, 232–249.
25. Smith, B.A.; Romo, R.V.; Smith, R.A.; de la Cruz, R.A.; Lime, B.J. Production of raw sugar from sorghum juices. *Sugar J.* **1973**, *35*, 22–27.
26. Rooney, W.L.; Blumenthan, J.; Bean, B.; Mullet, J.E. Designing sorghum as a dedicated bioenergy feedstock. *Biofuels Bioprod. Biorefin.* **2007**, *1*, 147–157.
27. Tew, T.L.; Cobill, R. Chapter 9: Genetic improvement of sugarcane (Saccharum spp.) as an energy crop. In *Genetic Improvement of Bioenergy Crops*; Vermerris, W., Ed.; Springer Science: Berlin, Germany, 2008; pp. 249–272.
28. Eggleston, G.; Heckemeyer, M.; St. Cyr, E.; Wartelle, L. Case study: Commercialization of sweet sorghum juice clarification for large-scale syrup manufacture. *Sugar Tech.* **2015**.
29. Fanjul, J.; Okeelanta Corporation, South Bay, FL, USA. Personal communication, 2014.
30. Serna-Saldivar, S.O.; Rooney, W.L. Chapter 8: Production and supply logistics of sweet sorghum as an energy feedstock. In *Sustainable Bioenergy Production*; Wang, L., Ed.; CRC Press: Boca Raton, FL, USA, 2014; pp. 193–212.
31. Dixon, T.F.; Bullock, G.E. Bioenergy outlook for Australia—For sugar not so sweet.*Zuckerindustrie* **2004**, *129*, 233–236.
32. Caio Filho, L. Bioelectricity: An increasing source of income in the sector. In *F. O. Licht 2010 Yearbook*; F. O. Licht: London, UK, 2010; pp. D13–D15.
33. Bagasse-based cogen sector expanding. Available online: https://www.internationalsugarjournal.com/brazil-bagasse-based-cogen-sector-expanding (accessed on 24 August 2015).
34. Castro, A.G.; Quiros, E. Cogeneration of energy from sugarcane bagasse in Costa Rica: The case of El Viejo factory. *Intern. Sugar J.* **1991**, *93*, 184–190.

35. Goldemberg, J.; Coelho, S.T.; Nastari, P.M.; Guardabasi, P. Chapter 9: Production and supply logistics of sugarcane as an energy feedstock. In *Sustainable Bioenergy Production*; Wang, L., Ed.; CRC Press: Boca Raton, FL, USA, 2014; pp. 213–222.
36. Fischer-Tropsch (FT). Available online: www.velocys.com (accessed on 24 August 2015).
37. Purchase, B.S.; Rosettenstein, S.; Bezuidenhoudt, D.V. Challenges and potential solutions for storage of large quantitites of bagasee for power generation. *Intern. Sugar J.* **2014**, *116*, 592–602.
38. Solomon, B.D.; Barnes, J.R.; Halvorsen, K.E. Grain and cellulosic ethanol: History, economics, and energy policy. *Biomass Bioenergy* **2007**, *31*, 416–425.
39. Petrou, E.C.; Pappis, C.P. Biofuels: A survey of pros and cons. *Energy Fuels* **2009**, *23*, 1055–1066.
40. Brennan, L.; Owende, P. Biofuels from microalgae—A review of technologies for production, process, and extraction of biofuels and co-products. *Renew. Sustain. Energy Rev.* **2010**, *14*, 557–577.
41. Dantas, G.A.; Legey, L.F.L.; Mazzone, A. Energy from sugarcane bagasse in Brazil: An assessment of the productivity and cost of different technological routes. *Renew. Sustain. Energy Rev.* **2013**, *21*, 356–364.
42. Yaman, S. Pyrolysis of biomass to produce fuels and chemical feedstocks. *Energy Conv. Manag.* **2004**, *45*, 651–671.
43. Spokas, K.A.; Cantrell, K.B.; Novak, J.M.; Archer, D.W.; Ippolito, J.A.; Collins, H.P.; Boateng, A.A.; Lima, I.M.; Lamb, M.C.; McAloon, A.J.; *et al.* Biochar: A synthesis of its agronomic impact beyond carbon sequestration. *J. Environ. Qual.* **2011**, *41*, 973–989.
44. Bapat, H.D.; Manahan, S.E. Chemchar gasification of hazardous wastes and mixed wastes on a biochar matrix. Available online: http://www.biochar-international.org/node/915 (accessed on 24 August 2015).
45. Mishra, B.K.; Ramakrishnan, P.S. Slash and burn agriculture at higher elevations in North-Eastern India. I. Sediment, water and nutrient losses. *Agric. Ecosyst. Environ.* **1983**, *9*, 69–82.
46. German, L.A. Historical contingencies in the coevolution of environment and livelihood: Contributions to the debate on Amazonian Black Earth. *Geoderma* **2003**, *111*, 307–331.

47. Mann, C.C. *1491: New revelations of the Americas before Columbus*; Vintage and Anchor Books: New York, NY, USA, 2005.
48. Bustin, R.M.; Guo, Y. Abrupt changes (jumps) in reflectance values and chemical compositions of artificial charcoals and inertinite in coals. *Int. J. Coal Geol.* **1999**, *38*, 237–260.
49. DeLuca, T.H.; MacKenzie, M.D.; Gundale, M.J. Biochar effects on soil nutrient transformations. In *Biochar for Environmental Management: Science and Technology*; Lehmann, J., Joseph, S., Eds.; Earthscan: London, UK, 2009; pp. 251–270.
50. Lehmann, J.; Rilig, M.C.; Thies, J.; Masiello, C.A.; Hockaday, W.C.; Crowley, D. Biochar effects on soil biota—A review. *Soil Biol. Biochem.* **2011**, *43*, 1812–1836.
51. Lehmann, J.; Gaunt, J.; Rondon, M. Bio-char sequestration in terrestrial ecosystems—A review. *Mitig. Adapt. Strat. Glob. Chang.* **2006**, *11*, 403–427.
52. Lima, I.M.; White, P.; Bigner, R.; Klasson, K.T. Economic products derived from sugarcane trash and bagasse. Available online: https://www.sugarjournal.com (accessed on 24 August 2015).
53. Zandersons, J.; Gravitis, J.; Kokorevics, A.; Zhurinsh, A.; Bikovens, O.; Tardenaka, A.; Spince, B. Studies of the Brazilian sugarcane bagasse carbonization process and product properties. *Biomass Bioenergy* **1999**, *17*, 209–219.
54. Pendyal, B.; Johns, M.M.; Marshall, W.E.; Ahmedna, M.; Rao, R.M. The effect of binders and agricultural by-products on physical and chemical properties of granular activated carbons.*Bioresour. Technol.* **1999**, *68*, 247–254.
55. Erlich, C.; Ohman, M.; Björnborn, E.; Fransson, T.H. Thermochemical characteristics of sugarcane bagasse pellets. *Fuel* **2005**, *84*, 569–575.
56. Cordero, T.; Marquez, F.; Rodriguez-Mirasol, J.; Rodriguez, J.J. Predicting heating values of lignocellulosics and carbonaceous materials from proximate analysis. *Fuel* **2001**, *80*, 1567–1571.
57. Channiwala, S.A.; Parikh, P.P. A unified correlation for estimating HHV of solid, liquid and gaseous fuels. *Fuel* **2002**, *81*, 1051–1063.
58. Garcia-Perez, M.; Chaala, A.; Roy, C. Vacuum pyrolysis of sugarcane bagasse.*J. Anal. Appl. Pyrol.* **2002**, *65*, 111–136.
59. Parikh, J.; Channiwala, S.A.; Ghosal, G.K. A correlation for calculating HHV from proximate analysis of solid fuels. *Fuel* **2005**, *84*, 487–494.

60. Munir, S.; Daood, S.S.; Nimmo, W.; Cunliffe, A.M.; Gibbs, B.M. Thermal analysis and devolatization kinetics of cotton stalk, sugarcane bagasse, and shea meal under nitrogen and air atmospheres. *Bioresour. Technol.* **2009**, *100*, 1413–1418.
61. Katyal, S.; Thambimuthu, K.; Valix, M. Carbonisation of bagasse in a fixed bed reactor: Influence of process variables on char yield and characteristics. *Renew. Energy* **2003**, *28*, 713–725.
62. Ahmedna, M.; Marshall, W.E.; Rao, R.M. Production of granular activated carbons from select agricultural by-products and evaluation of their physical, chemical and adsorption properties. *Bioresour. Technol.* **2000**, *71*, 113–123.
63. Ahmedna, M.; Marshall, W.E.; Rao, R.M. Surface properties of granular activated carbons from agricultural by-products and their effects on raw sugar decolorization. *Bioresour. Technol.* **2000**, *71*, 103–112.
64. Ng, C.; Bansode, R.R.; Marshall, W.E.; Losso, J.N.; Rao, R.M. Process description and product cost to manufacture sugarcane bagasse-based granular activated carbon. *Intern. Sugar J.* **2002**, *104*, 401–408.
65. Ng, C.; Marshall, W.; Rao, R.M.; Bansode, R.R.; Losso, J.N.; Portier, R.J. *Granular Activated Carbons from Agricultural By-products: Process Description and Estimated Cost of Production, Bulletin No. 881*; Louisiana State University AgCenter Research & Extension: Baton Rouge, LA, USA, 2003.
66. Johns, M.M.; Toles, C.A.; Marshall, W.E. Activated Carbons from Low-density Agricultural Waste. U.S. Patent 6,537,947, 25 March 2003.
67. Lehmann, J. Bio-energy in the black. *Front. Ecol. Environ.* **2007**, *5*, 381–387.
68. Bracmort, K.S. *Biochar: Examination of an Emerging Concept to Mitigate Climate Change*; CRC Report for Congress R40186; Congressional Research Service: Washington, DC, USA, 2009.
69. *The "Bagatex-20" Process for Rapid Drying of Baled Bagasse by Accelerated and Controlled Fermentation, Using a Biochemical Catalyst*; Groupe Sucres et Denrees: Paris, France, 1986; p. 49.
70. Springer, E.L. *Should Whole-tree Chips for Fuel be Dried before Storage?* Forest Products Laboratory Research Note FPL-0241; United States Department of Agriculture: Washington, DC, USA, 1979.
71. Wright, M.; Lima, I.M.; Bigner, R. Microbial and physicochemical properties of sugarcane bagasse for potential conversion to value-added products. *Sugar J.* **2015**, *78*. Article 31.

72. Parikh, J.; Channiwala, S.A.; Ghosal, G.K. A correlation for calculating elemental composition from proximate analysis of biomass materials. *Fuel* **2007**, *86*, 1710–1719.
73. Munoz-Arboleda, F.; Quintero-Duran, R. Trash management after green cane harvesting and its effect on productivity and soil respiration. *Intern. Sugar J.* **2011**, *113*, 804–808.
74. Viator, R.P.; Johnson, R.M.; Richard, E.P., Jr. Challenges of post-harvest residue retention management in the Louisiana sugarcane industry. *Proc. ISSCT Congr.* **2005**, *25*, 238–244.
75. Singh, P.; Suman, A.; Tiwari, P.; Arya, N.; Gaur, A.; Shrivatava, A.K. Biological pretreatment of sugarcane trash for its conversion to fermentable sugars. *World J. Microbiol. Biotechnol.* **2008**, *24*, 667–673.
76. Ripoli, M.L.C.; Ripoli, T.C.C.; Villa Nova, N.A. Sugarcane crop residues and bagasse for cogeneration in Brazil. *Am. Soc. Agric. Biol. Eng. Pap.* **2007**.
77. Muir, B.; Eggleston, G. Factory trials to determine how sugarcane trash impacts downstream processing including affinated sugar production. *Proc. Sugar Ind. Technol. Meet.* **2009**, *68*, 24–48.
78. Eggleston, G.; Grisham, M.; Tew, T.; Triche, R.; Antoine, A. Potential biomass quantity and processing quality of trash tissues by different US sugarcane varieties. *Intern. Sugar J.* **2009**, *111*, 108–118.
79. Deepchand, K. Chapter 4: The sugarcane crop for the sustainable production of sugar and other cane derived products in Mauritius. In *Sustainability of the Sugar and Sugar-Ethanol Industries*; ACS Symposium Series; Eggleston, G., Ed.; Oxford University Press: Oxford, UK, 2011; pp. 53–72.
80. Donaldson, R.A.; Redshaw, K.A.; Rhodes, R.; van Antwerpan, R. Season effects on productivity of some commercial South African sugarcane cultivars. Part II: Trash production. *Proc. S. Afr. Technol. Assoc.* **2008**, *81*, 528–538.
81. Eggleston, G.; Viator, R.; Gateuil, A.; Fenger, J.; White, P.; Jackson, W.; Waguespack, H., Jr.; White, P.; Blackwelder, N. Effects of seasonal variations of sugarcane stalk and extraneous matter quantity and quality on recoverable sugar, starch, and fiber (Part I). *Intern. Sugar J.* **2013**, *115*, 477–487.
82. Purchase, B.S.; Wynne, A.T.; Meyer, E.; van Antwerpen, R. Is there profit in cane trash?—Another dimension to the assessment of trashing versus burning. *Proc. S. Afr. Technol. Assoc.* **2008**, *81*, 86–99.

83. Schembri, M.G.; Hobson, P.A.; Paddock, R. The development of a prototype factory-based trash separation plant. *Proc. Aust. Sugar Cane Technol.* **2002**, *24*, 12–18.
84. Young, R.D. NFR Bioenergy to start $312 million sugar refinery project across South Louisiana. Available online: www.nola.com/business/baton-rouge/index.ssf/2014/09/nfr_bioenergy_to_start_312 (accessed on 24 August 2015).
85. Eggleston, G.; Viator, R.; Gateuil, A.; Fenger, J.; White, P.; Jackson, W.; Waguespack, H., Jr.; White, P.; Blackwelder, N. Seasonal variations of sugarcane stalk and extraneous matter on pH, color and ash as they affect the production of high quality raw sugars (Part II). *Intern. Sugar J.* **2013**, *117*, 646–653.
86. NFR BioEnergy LLC; White Castle, LA, USA. Personal communication, 2015.
87. Viator, R.P.; Richard, E.P., Jr.; Viator, B.J.; Jackson, W.; Waguespack, H.; Birkett, H. Sugarcane chopper harvester extractor fan and ground speed effects on sugar yield, cane quality, and field losses. *Appl. Eng. Agric.* **2006**, *23*, 31–34.
88. Eggleston, G.; Klich, M.; Antoine, A.; Beltz, S.; Viator, R. Brown and green sugarcane leaves as potential biomass: How they deteriorate under dry and wet storage conditions. *Ind. Crops Prod.* **2014**, *57*, 69–81.
89. Salassi, M.; Brown, K.; Hilbun, B.; Deliberto, M.; Gravois, K.; Mark, T.; Falconer, L. Farm-scale cost of producing perennial energy cane as a biofuel feedstock. *Bioenergy Res.* **2014**, *7*, 609–619.
90. Zhou, M.; Kimbeng, C.A.; Eggleston, G.; Viator, R.; Hale, A.; Gravois, K. Issues of starch in sugarcane processing and prospects of breeding for low starch content in sugarcane. *Sugar Cane Int.* **2008**, *26*, 3–13.
91. Ragauskas, A.J.; Williams, C.K.; Davison, B.H.; Britovsek, G.; Cairney, J.; Eckert, C.A.; Frederick, W.J., Jr.; Hallett, J.P.; Leak, D.J.; Liotta, C.L.; *et al.* The path forward for biofuels and biomaterials. *Science* **2007**, *311*, 484–489.
92. Alexander, A.G. *The Energy Cane Alternative*; Elsevier: New York, NY, USA, 1985.
93. Anonymous. Release of three high fiber sugarcane varieties: L 79-1002, Ho 00-961, and HoCP 91-552. *Sugar Bull.* **2007**, *85*, 21–26.

94. Doney, D.L.; Theurer, J.C. Potential of breeding for ethanol fuel in sugarbeet. *Crop Sci.* **1984**,*24*, 255–257.
95. Theurer, J.C.; Doney, D.L.; Smith, G.A.; Lewellen, R.T.; Hogaboam, G.J.; Bugbee, W.M.; Gallian, J.J. Potential ethanol production from sugar beeet and fodder beet. *Crop Sci.* **1987**,*27*, 1034–1040.
96. Libsack, S.; EnerGynetics, Minneapolis-St. Paul, MN, USA. Personal communication, 2014.
97. Buttazoni, M. *GHG Emission Reductions with Industrial Biotechnology*; World Wide Fund for a Living Planet Report; WWF: Copenhagen, Denmark, 2008.
98. Borghino, D. Artificial "Spiber" Silk is Tougher than Kevlar. Available online: http://www.gizmag.com/spiber-sythetic-sil/28267 (accessed on 24 August 2013).
99. Koch, T.J.; Venus, J.; Bruhns, M. Sugar beet syrups in lactic acid fermentation—Part I.*Zuckerindustrie* **2014**, *139*, 495–502.
100. Jacoby, M. Chemicals from the garden. C&E News, 6 July 2009, 26–28.
101. Bomgardner, M.M. Biobased polymers. C&E News, 27 October 2014, 10–14.
102. Werpy, T.; Petersen, G. Top Value Added Chemicals from Biomass: Results from Screening for Potential Candidates from Sugars and Synthesis Gas. Available online: www1.eere.energy.gov/biomass/pdfs/35523.pdf (accessed on 24 August 2004).
103. Scott, A. Biomaterial blues. C&E News, 20 October 2014, 20–21.
104. Tullo, A.H. Endgame nears for Metabolix. C&E News, 25 May 2015, 26–27.
105. Godshall, M.A. Future directions for the sugar industry. *Intern. Sugar J.* **2001**, *103*, 378–384.
106. Walford, S.N.; Morel du Boil, P.G. A survey of value addition in the sugar industry. *Proc. S. Afr. Technol. Assoc.* **2006**, *80*, 39–611.
107. Eggleston, G.; Salassi, M.; Richard, E.; Birkett, H. Sustainability of the sugar industry: Future value addition from sugarcane. *Intern. Sugar J.* **2007**, *109*, 415–432.
108. Godshall, M.A. Value-added products for a sustainable sugar industry. In *Sustainability of the Sugar and Sugar-Ethanol Industries*; ACS Symposium Series; Eggleston, G., Ed.; Oxford University Press: Oxford, UK, 2010; pp. 253–268.

109. Kralisch, D. Applications of LCA in process development. In *Green Chemistry Metrics*; Lapkin, A., Constable, D., Eds.; Wiley & Sons: Oxford, UK, 2008; pp. 248–271.
110. Ritter, S.K. Common ground for going green. Chem. Eng. News, 10 May 2010, 38–41.
111. Rosenzweig, M. Group offers sustainability guide. Available online: http://www.chemicalprocessing.com (accessed on 24 August 2015).
112. Ottewell, S. Sustainability Efforts measure up. Chemical Processing, 27 April 2015, 20–25.
113. Rein, P. Measuring and monitoring sustainability in the sugar and sugar-ethanol industry. In *Sustainability of the Sugar and Sugar-Ethanol Industries*; ACS Symposium Series 1058; Eggleston, G., Ed.; Oxford University Press: Oxford, UK, 2010; pp. 23–35.
114. Trembley, J.-F. Sustainability yields sweet success. C&E News, 1 June 2015, 18–19.
115. Hart, S.L. Beyond greening strategies for a sustainable world. In *Harvard Business Review on Green Business Strategy*; Harvard Business School Publishing Corporation: Watertoen, MA, USA, 2007; pp. 99–123.
116. Leale, L.L. Industry is reducing its ecological footprint. Food Quality and Safety, 15 June 2015, 20–25.

6
CHAPTER

DESIGN OF SUSTAINABLE BIOFUEL PROCESSES AND SUPPLY CHAINS: CHALLENGES AND OPPORTUNITIES

George G. Zaimes[1], Nemi Vora[1], Shauhrat S. Chopra[1], Amy E. Landis 2 and Vikas Khanna [1,*]

[1]*Swanson School of Engineering, Department of Civil and Environmental Engineering, University of Pittsburgh, Pittsburgh, PA 15621, USA*

[2]*Glenn Department of Civil Engineering, Clemson University, Clemson, SC 29631, USA*

ABSTRACT

The current methodological approach for developing sustainable biofuel processes and supply chains is flawed. Life cycle principles are often retrospectively incorporated in the design phase resulting in incremental environmental improvement rather than selection of fuel pathways that minimize environmental impacts across the life cycle. Further, designing sustainable biofuel supply chains requires joint consideration of economic, environmental, and social factors that span multiple spatial and temporal scales. However, traditional life cycle assessment (LCA) ignores economic aspects and the role of ecological goods and services in supply chains, and hence is limited in its ability for guiding decision-making among alternatives—often resulting in sub-optimal solutions. Simultaneously incorporating economic and environment objectives in the design and optimization of emerging biofuel supply chains requires a radical new paradigm. This work discusses key research opportunities and challenges in the design of emerging biofuel supply chains and provides a high-level overview of the current "state of the art" in environmental sustainability assessment of biofuel production. Additionally, a bibliometric analysis of over 20,000 biofuel research articles from 2000-to-present is performed to identify active topical areas of research in the biofuel literature, quantify the relative strength of connections between various biofuels research domains, and determine any potential research gaps.

Keywords:biofuels; sustainability; sustainable biofuel supply chain; next generation biofuels; bibliometric analysis; multiobjective optimization; multiscale modeling; life cycle assessment

1. INTRODUCTION

Chemical technologies and the chemical process industry provide a range of useful and valuable products derived from biobased resources for use in personal care products, health products, agrochemicals, and transportation fuels. However, the production of these products is accompanied by generation of vast quantities of wastes and a range of harmful emissions to air, water, and soil. There is increasing realization that resource consumption and anthropogenic-derived impacts can have long-standing consequences on global ecological systems, and place strain on the natural biogeochemical cycles that support human life. Published findings from the millennium ecosystem assessment (MEA)—an international collaboration designed to assess the impact and widespread consequences of environmental change for human and ecological well-being, indicate that in the second half of the 20th century anthropogenic-derived resource degradation and overconsumption of natural capital have changed ecosystems more rapidly and extensively than in any comparable period in human history [1]. Rockstrom and colleagues assessed that the Earth has transgressed planetary boundaries for climate change, biodiversity, nitrogen cycle balance and is fast approaching the limit of safe operating space for global freshwater use, land use change, ocean acidification and global phosphorous cycle balance [2,3,4,5].

Traditional methods of chemical process design have primarily relied on finding the economic optimum subject to physical constraints, namely satisfying the heat and material balances and thermodynamic limitations. However, concerns over the depleting fossil energy sources, mounting regulatory compliance and the resulting push towards environmentally conscious process design are forcing designers to consider reduced environmental impact as one of the product design objectives. Business leaders have begun to realize that such a shift towards more sustainable design practices can not only minimize the environmental impact of industrial activity but is also crucial for long term success and sustainability of their enterprises.

The emerging field of sustainability science and engineering is developing tools to recognize, quantify, and reconcile resource limitations; human needs, and optimize global and human benefit. The concept of sustainability is multifaceted; encompassing the entirety of the human enterprise, interfacing with environmental, social, political, and economic issues, and as such is highly interdisciplinary. The outstanding challenge facing the chemical industry is the incorporation of environmental and sustainability objectives along with traditional design objectives in the development of emerging chemical processes. The rapid development of biofuels as a potentially sustainable and cleaner replacement for conventional fuels represents a unique challenge for the chemical industry that requires simultaneous consideration of economic, social, and ecological aspects and thus exemplifies an excellent content in which to understand the challenges and opportunities for designing sustainable supply chains.

2. EMERGING BIOFUEL PATHWAYS

Global climate change, volatile petroleum prices, energy security issues, and resource depletion have driven nations to consider adding renewable and alternative energy options to their energy portfolios. Transportation as well as the power and electricity generation sector constitute the two largest greenhouse gas (GHG) emitting sectors in the United States, accounting for 27% and 31% of total GHG emitted in 2013 respectively [6]. Accordingly, this has prompted the development of United States (U.S.) regulatory programs as well as European directives designed to mitigate GHG emissions from the transportation and power generation sector while concurrently increasing domestic energy independence and security. In 2007 the U.S. congress passed the Energy Independence and Security Act (EISA) [7], which mandates the production of 36 billion gallons of biofuels by the year 2022, and stipulates that a percentage must be derived from conventional, cellulosic biofuel, biomass-based diesel, and advanced biofuels. Additionally, EISA sets yearly volumetric biofuel production targets and requires renewable fuels achieve minimum reductions in overall life cycle GHG emissions relative to baseline petroleum fuels. Similarly, in

2009 the European Union (EU) passed the renewable energy directive (RED) [8]. This legislation requires that the EU obtain 20% of its total energy consumption from renewable sources, and derive 10% of energy consumption within the transportation sector from renewable resources by 2020. Furthermore, the EU imposed a 5% cap on the amount of food crop-derived biofuels used to meet the EU's 2020 goal, in an effort to mitigate the potential social and economic impacts of competition between crops for food *vs.* fuel. The RED sets minimum life cycle GHG emission reductions targets of 35% relative to baseline petroleum fuels for the year 2010, increasing to 50% in 2017 and 60% in 2018.

Currently ethanol from corn is the most widely produced and utilized biofuel in the United States. However, critical concerns have been raised about the potential of corn ethanol and other first generation biofuels in mitigating climate change and reducing dependence on fossil fuels. It has been contended that the direct and indirect land use change effects may possibly negate the GHG reduction potential of first generation biofuels possibly resulting in overall higher life cycle GHG emissions relative to baseline petroleum fuels [9,10,11]. In addition, first generation biofuels require changes in the existing transportation infrastructure such as modifications to vehicle engines and fuel pipelines. Further, the production and use of first generation biofuels have resulted in deleterious impacts on ecosystem goods and services such as soil erosion, water and air pollution, and loss of biodiversity, and thus have prompted the development of next generation biofuels [12].

A myriad of feedstocks and conversion platforms pathways are currently under development for next generation biofuel production, shown in Figure 1. Common biofeedstocks include sugarcane and corn, lignocellulosic biomass, algae, and oil-seeds. Lignocellulosic feedstocks include agricultural residues such as corn stover and forest residues; energy crops such as switchgrass, *Miscanthus*, and poplar; and industrial/municipal solid wastes. Additionally, several non-food lipid sources including algae and *Jatropha* are being considered for biofuel production. The "Billion Ton Study" jointly conducted by the US Department of Energy (DOE) and the US Department of Agriculture (USDA) provides a detailed quantification of available biomass in the US for producing fuels and biobased products, and estimates that up to 1.6 billion tons of dry biomass can be produced annually in the contiguous U.S. for meeting growing energy demands without displacing critical food, fiber, or feed crops [13]. These findings are compelling as they suggest that the U.S. has the capacity to support a large-scale domestic biofuels and bioproducts market.

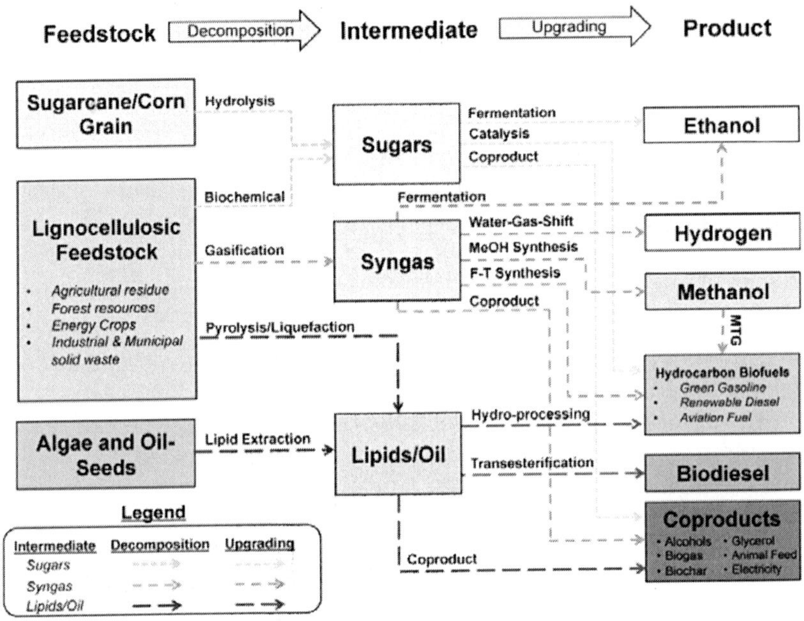

FIGURE 1. Feedstocks and conversion pathways for biofuel production. MTG: Methanol to Gasoline. F-T Synthesis: Fischer-Tropsch Synthesis.

Several comprehensive reviews of the possible catalytic, biochemical, and thermochemical conversion pathways for the production of biofuels from biomass feedstocks are available [14,15,16,17]. Traditionally, fats and oils derived from oilseed crops are converted to biodiesel via transesterification involving chemical reactions with an alcohol, such as methanol. However, biodiesel has certain undesirable properties such as poor cold flow properties, high cloud point, and roughly 10% lower energy density when compared with petroleum derived diesel fuel. Alternatively, oils and fats can be catalytically hydrotreated and converted to renewable diesel—a renewable hydrocarbon fuel that is fungible with traditional petroleum fuels. Lignocellulosic feedstocks, such as woody biomass and agricultural residues, require acid treatments to decompose cellulose and hemicellulose into sugars, which can be subsequently reformed into hydrocarbons via microbial agents [18]. Dumesic and coworkers have unraveled an entire set of catalytic reactions that convert biomass into hydrocarbon biofuels in the gasoline, diesel, and jet range [16,19,20,21,22,23,24,25]. Bond and coworkers successfully reported a strategy for converting γ-valerolactone, an intermediate produced from biomass carbohydrates, into liquid alkenes targeted for transportation fuels [26]. Their proposed integrated catalytic system eliminates the need for an external source of hydrogen.

Thermochemical conversion via pyrolysis or gasification has gained widespread attention as commercial platforms for biofuel production [27]. Pyrolysis involves thermochemical decomposition of organic matter at 400–600 °C in absence of oxygen and produces oxygenated bio-oil as well as biochar and non-condensable gases as coproducts [28,29,30,31,32]. Pyrolysis oil can be catalytically converted to high-octane hydrocarbon biofuels in the gasoline and diesel range [29,33,34], while several uses for the bio-char including (1) a soil amendment or (2) combustion to produce electricity are actively being investigated [35,36,37,38,39,40,41]. Agrawal and Singh presented a conceptual design of a hydrogen bio-oil process using fast pyrolysis and hydrodeoxygenation for producing liquid transportation fuels that require only modest quantities of supplementary H_2 [42]. Another promising thermochemical pathway is gasification, in which biomass is converted to synthesis gas (syngas)—a mixture containing CO, H_2, CH_4, N_2 and CO_2 by heating the input feed to high temperatures (>700 °C) and reacting with air, oxygen, and/or steam. Syngas can then be converted into liquid fuels by well-known reactions such as Fischer-Tropsch synthesis or synthesizing methanol and subsequent conversion to gasoline. Dauenhauer and coworkers at the University of Minnesota have combined the biomass gasification, tar cleanup, and water-gas shift into a single reactor [43].

Recently, microalgae have been touted as a promising feedstock for conversion to liquid transportation fuel due to their unique ability to utilize carbon dioxide from industrial activities [44], high photosynthetic efficiency [45], and ability to be grown on marginal or otherwise non-arable land [46]. Algae derived biofuels do not directly displace or put market pressure on food crops, as do the production of biofuels derived from corn, soybean, or sugarcane. Furthermore, algae can be grown using different growth media including wastewater as well as brackish/saline water; the use of which has the capacity to substantially reduce the nutrient and water-footprint of microalgal fuels relative to traditional terrestrial biofeedstocks [47]. Algae can be grown in a semi-continuous to continuous manner, and thus do not require perennial harvesting such as other leading forms of biomass including poplar or perennial grasses. Brennan and Owende presented a comprehensive overview of the different technologies for microalgae cultivation, harvesting, conversion technologies, and the potential useful products including biofuels [48]. Furthermore, an emerging concept of using cyanobacteria to produce ethanol through intracellular photosynthetic process in cyanobacteria has shown promise due to its carbon footprint reduction potential [49].

Anaerobic digestion has gained interest as a process for producing renewable energy from biomass resources. Anaerobic digestion is a biological process in which microorganisms decompose biodegradable material, in the absence of oxygen, to produce biogas that is comprised primarily of carbon dioxide and methane. Biogas can be further upgraded via Fischer-Tropsh synthesis, or can be used for direct combustion for space heating or electricity production after further upgrading. Anaerobic digestion is often implemented in emerging microalgal biofuel systems [50], and has been cited as a crucial process for enhancing the environmental sustainability of microalgae to biodiesel systems as it can be used to recover biogas from micro-algal residue [51]. Another alternative to conventional biomass is use of a combination of perennial low input high diversity (LIHD) crops. These crops require very little input in terms of water, fertilizer and nutrients and have the ability to grow on degraded land; mitigating issues of food *vs.* fuel and boosting regional biodiversity. Additionally, these benefits enable LIHD based biofuel systems to produce higher net energy output per fossil fuel input compared to conventional biofuels [52]. However, due to lower biomass yields, LIHD biomass is often only used in decentralized, small-scale combustion and anaerobic digestion conversion systems [53,54,55]. The biofuel research landscape is rapidly evolving in regards to process multiplicity, complexity, and scope. A comprehensive review of historical trends in biofuel production and policy are beyond the scope of the current work but can be found in several notable studies [56,57,58]. Sustainable commercialization of biofuel production will require simultaneously addressing multiple technical, economic, social, and environmental challenges that occur throughout the supply chain, and thus inherently necessitates a collaborative approach.

3. BIBLIOMETRIC ANALYSIS OF BIOFUEL LITERATURE

Understanding the full range of potential environmental, social, and economic impacts of biofuel production prior to its widespread commercialization and use is pivotal for avoiding unintended consequences and for guiding the sustainable development of the biofuels industry [59]. Moreover, holistic assessment of the widespread direct and indirect impacts of biofuel production requires integrating data and information across multiple research domains and rigorous analysis of peer-reviewed literature. As such, collaboration and synthesis across multiple disciplines is necessary for environmentally conscious decision-making. However the degree to which interdisciplinary research is occurring between biofuel research domains is often not well-understood or studied, such information is critical for determining research gaps, fragmentation between research domains, and for guiding the future trajectory of research in biofuels.

We performed a bibliometric analysis to: (1) identify active topical areas of research in the biofuel literature; and (2) quantify the relative strength of connection between various biofuels research domains via analyzing the occurrence and co-occurrence frequency of author supplied and indexed keywords from over 20,000 biofuels articles published from 2000 to present. The resulting analysis provides useful insights regarding the strength of connection and coupling between various research domains in the biofuel literature, as well as potential research gaps, *i.e.*, research areas that may require further synthesis and integration. Keywords for over 20,700 articles were obtained using the search-term "Biofuels" from the SCOPUS database [60]. In absence of author-supplied keywords, indexed keywords were used. This approach was leveraged over using both author-supplied keywords and indexed keywords, so as to avoid potential double counting. Keywords were aggregated into 36 topic areas related to 7 broad themes; shown in Table 1. The analysis expands on the method of identifying and aggregating keywords provided in Ridley *et al.* [61], please see supporting information for a complete list of keywords and method of data aggregation. The analytic framework and bibliometric algorithm established in Van Eck and Waltman was used to detect the number of co-occurrences of select keywords [62]. The co-occurrence data was mapped via network software (ORA) [63] to visualize the dynamic interactions between select biofuel research domains. The strength of the linkage between research topics (nodes) indicates its relative co-occurrence and is proportional to the line width, while the size of the node indicates its relative occurrence.

Figure 2 presents a co-occurrence network graph, illustrating the connectivity between various research topics in the biofuels literature. Figure 2 reveals that greenhouse gases, biodiesel, ethanol, transesterification, and hydrolysis had the highest frequency of co-occurrence in the literature, while market forces, trade, biodiversity, ecosystems goods and services (EGS), thermodynamics, aviation fuel, and hydrocarbon biofuels were not well represented. Nodes with no connections represent research fields that lack significant collaborations with other research areas in the biofuel literature, and may be emerging or nascent topics. Figure 2 reveals that substantial research has been invested in evaluating GHG emissions related to biofuel production, while little effort has been made to evaluate impacts on soil resources, human health, food security, biodiversity, and ecological goods and services. While GHG emissions represent an important sustainability aspect for biofuel production, other categories must also be considered so that biofuels do not inadvertently shift their impacts to other domains.

TABLE 1. Themes and topics for author supplied and indexed keywords. A detailed list of all keywords is in Table S1.

Theme	Topic	Example Keywords
Environmental & Human Wellbeing	Food Security	Food Supply, Food Crop
	Human Health	Mortality, Asthma
	GHGs	Greenhouse Gases, Carbon Dioxide
	Air Quality (Non-GHGs)	Particulate Matter, Volatile Organic Compounds
	Soil Resources	Soil Organic Carbon, Soil Fertility
	Land Use Change	Indirect Land Use Change, Direct Land Use Change
	Water Resources	Groundwater, Water Footprint
	Biodiversity	Wildlife, Biodiversity
	Life Cycle Assessment	Life Cycle Analysis, Life Cycle Assessment
	Ecosystem Services	Ecosystem Services, Ecosystems
Economy	Cost of Production	Technoeconomic analysis, Infrastructure
	Market Forces	Supply and Demand, Cost Competitiveness
	Policy	RFS2, EISA, LCFS
	Trade	Import, Export, Tariff
Conversion Platforms	Pyrolysis	Fast Pyrolysis, Pyrolysis oil
	Gasification	Gasification, BTL
	Transesterification	Esterification, FAME
	Hydrolysis	Hydrolysis, Fermentation
	Anaerobic Digestion	Anaerobic Digestion
Feedstocks	Woody Biomass	Willow, Poplar
	Perennial Grasses	Switchgrass, *Miscanthus*
	Oil Seeds	*Jatropha*, Soybean, Rapeseed
	Algae	Microaglae, Macroalgae
Feedstocks	Agricultural Residue	Sugarcane Bagasse, Corn Stover, Forest Residue
	Industrial & Municipal Waste	Waste Cooking Oil, Vegetable Oil
	Grains & Sugar Crops	Corn, Wheat, Rye
Thermodynamics	N/A	Exergy, Emergy

Adapted from Ridley *et al.* [61].

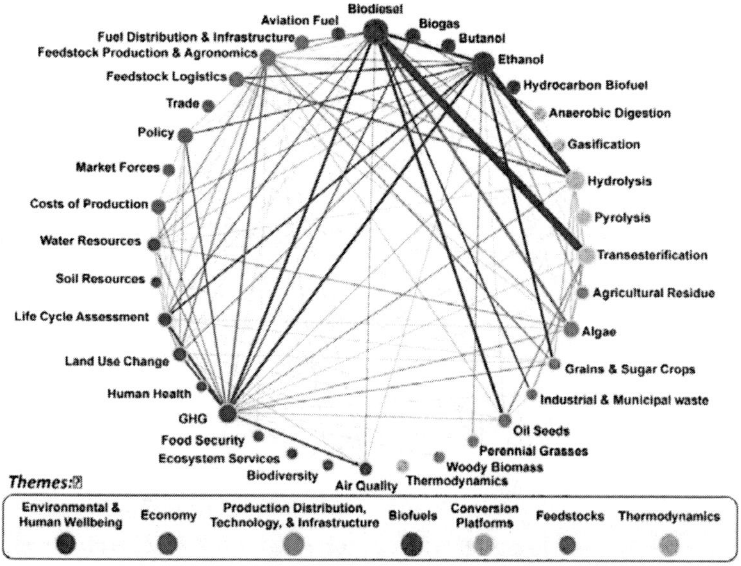

FIGURE 2. Bibliometric network analysis of biofuel research articles published between 2000 to present. The size of the nodes is scaled up as per the number of times each keyword occurs in the search. The width of the edges represents the co-occurrence of specific keywords. Colors of the nodes represent distinct research topic areas. Note: a co-occurrence cutoff value of 80 was chosen to identify significant connections between nodes in the network; as such some nodes appear to have no connections.

Furthermore, the lack of research (represented by lack of connecting links) on ecological goods and services, biodiversity, soil resources, and water resources is alarming, as prior research has suggested that the adoption of first generation biofuels have resulted in higher rates of deforestation and soil erosion, loss of biodiversity, and increased stress on water resources [64,65]. Hydrocarbon biofuels as well as aviation fuels were found to have marginal to no connectivity with other research domains, as these biofuels represent emerging topics and have only recently received widespread scientific inquiry. As such, further collaborative research is needed in all aspects of these emerging fuel platforms. The network diagram shows sparse scientific coverage on the market forces and/or trade aspects of biofuel production, despite consistent growth in the international trade of biomass/biofuels over the past decade. Failure to consider the broader trade and market implications of biofuel production could have significant global economic and social repercussions. This is particularly important for developing countries that are increasing biofuel exports to meet international policy mandated volumetric biofuel production and renewable energy targets, as developing nations are often highly sensitive to the potential adverse impacts of biofuel production including accelerated destruction of natural ecosystems, agricultural runoff and soil erosion, increased food insecurity and malnutrition, and global climate change. The results of the bibliographic analysis highlight the need for further interdisciplinary research so to assess any potential dynamic interactions, feedbacks, trade-offs, and mitigate any unintended consequences of biofuel production.

4. DESIGNING SUSTAINABLE BIOFUEL SUPPLY CHAINS

The process of introducing and implementing environmentally benign strategies in chemical process design started with the introduction of heat integration strategies. These schemes were primarily introduced in response to the energy crisis at that time and not only led to reduction in the overall energy consumption in production plants but also increased revenues by minimizing the plant operating costs. Although these strategies introduced the concept of environmentally conscious design, they generally had a narrow scope and boundary. Traditionally, the chemical process industry has responded to the environmental challenge by resorting to end-of-pipe solutions or pollution remediation such as recycling, waste treatment, and disposal. Although these practices are valuable, they may require vast capital and operating expenditures and can often shift the domain of pollution by moving it outside the analysis boundary. It is for this reason that in the last decade strategies and concepts such as green chemistry, life cycle assessment, environmentally conscious process design, and design for environment have come into prominence [66,67]. These concepts are attractive with their primary focus on avoiding waste generation rather than waste minimization. The unifying theme behind these approaches is to look at systems holistically by expanding the system boundary in traditional process design. Although, expanding the system boundary for a more holistic analysis is appealing, it presents a new challenge associated with the need for increased amount of data and computational time. With appreciation of such challenges and to address the shortcomings of some of these earlier approaches, process systems engineering has gradually expanded its analysis boundary from the narrowly focused process scale to the life cycle or supply chain scale and more recently to the ecosystem scale [68,69,70,71,72]. These approaches take a more holistic view by focusing on the entire life cycle or supply and demand webs of the selected products or processes. Zhuang *et al.* reviewed several existing modeling approaches for sustainable chemical production and their applications at multiple scales ranging from metabolism, to the life cycle, to ecosystems, and proposed a multi-scale approach integrating these models into a single cohesive framework [73].

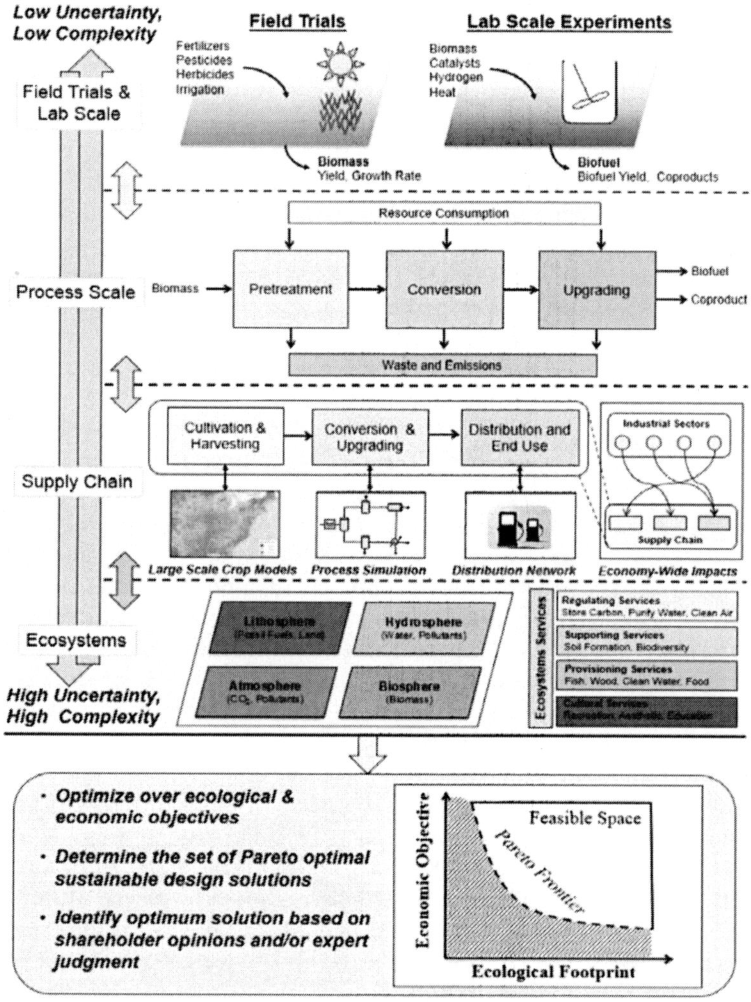

FIGURE 3. Modular Multi-scale, Multi-objective, Biofuel Supply Chain Optimization Framework.

Recent interest in biofuels has led to the development and use of models and computational tools at multiple scales including large-scale crop models, detailed chemical process design simulations, life cycle assessment models, and mathematical optimization tools. While these computational methods each provide unique and novel insights into the sustainability of emerging biofuels, these tools are often used in isolation and thus are limited in their ability for guiding decision-making. Synthesis of these models and tools into a unified framework, via collaboration between researchers across disciplines and modeling scales is required to provide a broader understanding of the sustainability of emerging biomass-to-fuel supply chains. Accordingly, this work discusses a modular multi-scale and multi-objective framework spanning from the field/lab scale, to the detailed process scale, the life cycle scale, and finally the ecosystems scale for holistic sustainability assessment of biofuel production; see Figure 3. The envisioned multi-scale approach evaluates the process in a hierarchical fashion, starting from the field/lab scale and expanding the system boundaries as successive scales are added. Information from lab/field trials such as reactor kinetic studies, pilot-scale biomass growth trials, and experimental trials on biofuel yields are used to parameterize design blocks and crop models used at the process level. Information such as liquid product distribution and operating plant utility requirements obtained via the process level is subsequently utilized to model unit processes in the supply chain. Information at the supply chain is coupled with the larger economy and ecosystems via the use of environmentally extended economic models. The proposed framework synthesizes results obtained from models/methods across disciplines and scales (*i.e.*, the lab/field scale, process scale, supply chain, and ecosystem scale) to inform process design and decision-making. Insights obtained from models/methods at differing scales (as well as their limitations) are disseminated between researchers working at different modeling tiers. This multi-scale interdisciplinary approach provides stakeholders different tiers of decision-making criteria (*i.e.*, capital and operating costs, environmental damages, or ecological impacts), and thus a holistic understanding of the broader consequences of emerging fuel pathways. Further, such an approach is conceptually attractive since it facilitates the evaluation procedure starting with simple systems and increasing complexity gradually as successive information layers are added. This approach can allow for screening out bad alternatives, for example, those with a negative economic potential early in the design stage thus saving computational time and providing a range of alternatives to the decision maker while avoiding arbitrary combinations. Further, the proposed framework considers multi-objective optimization over the broader superstructure to identify supply chain configurations that optimize ecological and economic performance while simultaneous achieving minimum threshold sustainability criteria. Design opportunities and challenges for each scale of analysis are discussed in the following sections.

4.1. Field Trials and Laboratory Scale Experiments

The first level of analysis involves field trials and laboratory scale experiments such as model compound studies [74,75], effects of catalysts and reactor conditions on product streams [76], estimation of fuel properties [77], reactor kinetic studies [78,79], pilot-scale biomass growth trials [80], and the effect of varying fertilizer and management practice on biomass yield [81,82]. This research is often focused on understanding the mechanism or principles underlying experimental observation, or determining the technically feasibility of an emerging technology. Factorial designs are often implemented to study the effect of each effect factor on the response variable (*i.e.*, biofuel yield), as well as the effect of interactions between factors on the response variable [83]. Further, data at the field/lab scale typically has low uncertainty, and is often used for calibration and parameterization in process models.

4.2. Process Scale

The next tier of analysis involves the use of agricultural crop models as well as traditional process design analysis. A variety of large-scale crop models have been developed to simulate bioenergy crop production including herbaceous crops (e.g., EPIC, ALMANAC, MISCANMOD, MISCANFOR, WIMOVAC, Agro-IBIS, Agro-BGC, APSIM, AUSCANE, LPJmL, CANEGRO), woody bioenergy crops (e.g., 3PG, SECRETS), and crassulacean acid metabolism crops (e.g., EPI) [84]. These models simulate biomass yield, nutrient cycling, water requirements, carbon flux, and other key parameters under different crop management practices. Recent efforts have been made to integrate these models with Geographic Information System (GIS) to create spatially explicit large-scale crop models [85]. Conversion of biomass to fuel can be modeled using conventional process simulators such as Aspen Plus, ChemCAD, or SuperPro [86]. Inputs to the biorefinery generally include raw materials and utilities such as biomass, steam, electricity, and cooling water. Similarly, final products typically include biofuel, coproducts, waste, and emissions. Process simulators provide a scientifically rigorous method for determining the utility requirements as well as material and energy flows for conversion of biomass to renewable transportation fuel. Information at this scale of analysis is often used for developing technoeconomic models [30,87], with the primary objective of maximizing material and energy efficiency while concurrently minimizing operating costs. Life cycle considerations such as the embodied impacts of material and energy inputs are not considered within the scope of analysis at this scale, thus decision-making based solely on information at the process scale could result in unsustainable design choices. For example, analysis based solely on the process scale could result in selection of high quality resources (*i.e.*, fossil fuels) for maximizing plant performance and economics; however, these resources often have high upstream environmental or human health impacts.

4.3. Modeling the Supply Chain and Life Cycle

The third level of analysis extends the analytic boundary to consider the material, energy, and emissions flows throughout the entire supply chain. This holistic systems approach captures environmental impacts that are outside the purview of the traditional process design boundary. Life Cycle Assessment (LCA) is one of the most common approaches for evaluating the environmental impact of a product or a process over its entire life cycle, and in recent years has emerged as the predominant method for analyzing the environmental sustainability of emerging biofuel platforms [88,89,90,91,92,93,94,95]. LCA considers impacts throughout all stages of the fuel life cycle—from raw material acquisition, to fuel conversion, and final use. LCA allows for a comprehensive understanding of the environmental impacts that occur at each stage of the supply chain, enabling the LCA practitioner to identify processes responsible for highest environmental burden and thus target these areas for process improvement. LCA can be used to quantify the anticipated impacts of a product or service prior to its widespread adoption, thus identifying and avoiding potential environmental pollutants, wastes, and environmental damages before they become embedded within the supply chain. Further, LCA can be used to compare the environmental performance between two products with the same functionality and can inform environmentally conscious decision-making. Garcia and You reviewed major challenges and opportunities in supply chain design and optimization—identifying several key technical challenges including: (i) multiscale challenges; (ii) multiobjective and sustainability challenges; and (iii) multi-player challenges [96]

4.3.1. Process LCA

LCA is a data intensive approach and has been extensively applied to study biofuel systems over the past decade. Several different LCA modeling methods have been developed; the most widely used LCA approach (*i.e.*, Process-LCA) defines a finite boundary by selecting the most important processes in a life cycle [97]. Data concerning the resource consumption and emissions for these processes are developed and compiled to generate a life cycle inventory (LCI). The life cycle impact assessment (LCIA) phase translates the energy, resource, and emissions flows identified in the LCI into their potential consequences for human health and the environment, and consists of a two-step process of impact classification and quantitative characterization. The classification step links each LCI flow with its related impacts on resource use, human health, and the environment. The characterization step calculates the magnitude of the associated impacts in terms of a reference unit for each category via multiplying the related resource, material, or energy flows with their respective impact factors. Translating the environmental impacts to a reference unit provides a common basis or measure for the generated impact, so that different emissions and resources can be compared and aggregated using a common unit. Data required for LCA can be obtained via commercial life cycle databases (e.g., ecoinvent [98]), publicly available life cycle data (e.g., OpenLCA [99], USLCI [100], GREET [101]), information from the open literature, or proprietary information.

Although widely used, process LCA suffers from many limitations including the use of an arbitrary life cycle boundary, combining data in disparate units and at multiple spatial and temporal scales, dealing with high dimensionality data involving varying degrees of uncertainty, and dealing with processes having a range of emissions [11]. Furthermore, for a system that simultaneously produces multiple products and coproducts these is no universally accepted method as to how to apportion the environmental impacts amongst said products. This is particularly important for biofuel systems in which non-fuel coproducts represent a significant fraction of total market value, mass, or energy flow as the choice of allocation procedure can often yield divergent results concerning the sustainability of these systems [102,103]. Additionally, biofuel LCAs often utilize differing functional units, system boundaries, allocation schemes, impact assessment methods, and report different sustainability metrics. Consequently, it is not unusual for LCA practitioners to obtain contradictory LCA results for the same system; this discrepancy has led to several harmonization and meta-analysis studies in the biofuels literature [104,105]. Data used in process LCA is at an intermediate scale since it is typically averaged to represent manufacturing processes, thus making it of limited use for making environmentally conscious engineering decisions about an individual process or equipment, which are at a finer scale, or for evaluating the effect on the macro economy, which is at a coarser scale.

4.3.2. EIO-LCA and Hybrid LCA

Input-output (IO) models, first developed by Nobel Prize winning economist Wassily Leontief, provide a mathematical framework for quantifying the inter-industrial connections and economic flows between different industrial sectors in the economy [106]. The traditional IO framework can be extended to consider the environmental impacts, emissions, and resource use for industrial sectors in the economy; and thus be utilized to perform LCA at the economy scale. This approach, known as Economic Input-Output LCA (EIO-LCA) [107], does not suffer from the challenge of defining a finite life cycle boundary as does Process LCA. Further, EIO-LCA uses a relatively complete network, but at a coarse scale of resolution. Data at this scale often do not include the use phase of the life cycle. Recently, Bakshi and colleagues have developed Ecologically-based Life Cycle Assessment (EcoLCA), an environmentally extended input-output life cycle model capable of accounting for the consumption/role of ecosystem goods and services in a life cycle framework [108,109]. The EcoLCA model extends the traditional I-O framework to consider the direct and indirect environmental impacts that result from economic activities; including ecological and natural resource consumption, emissions, land-use, and other environmental impact categories [110,111,112]. EcoLCA quantifies ecological resource consumption using a hierarchy of thermodynamic metrics including energy, industrial cumulative exergy consumption (ICEC), and ecological cumulative exergy consumption (ECEC), as well as mass flow. However, while exergy-based methods for thermodynamic aggregation of natural resource consumption may provide useful insights, these methods have their own limitations and are debated in the literature [113,114,115,116,117,118,119,120]. Research on combining the best advantages of Process LCA and EIO-LCA has also resulted in Hybrid LCA approaches, which combine the details of Process LCA with the greater completeness of EIOLCA [121,122]. Hybrid LCA attempts to balance computational tractability, completeness, and the use of detailed information.

4.3.3. Attributional *vs.* Consequential LCA

LCA models may also differ in the approach employed to address the material and energy flows in the system under investigation. The attributional LCA (ALCA) methodology, which has been utilized for a vast majority of the LCA studies, attempts to quantify the flow of resources and emissions from a product system and its subsystems. Emissions and their impact are attributed to the final product by one of the several available methods (allocation or system expansion). However, researchers have argued that it is not fully possible to draw conclusions on future changes by using only ALCA [123,124]. In contrast, consequential LCA (CLCA) methodology, aims to explain how the physical flows to and from the technosphere may change in response to a change in the life cycle of the product or service [125]. CLCAs attempt to consider a much broader system boundary. The most commonly employed form of a CLCA considers the use of economic models that track monetary, material, and energy flows across economic systems. This is generally accomplished using marginal data and is accounted for on the basis of price elasticity of supply and demand [126,127,128].

The US Environmental Protection Agency (US EPA) has used a multi-market multi-region partial equilibrium model to determine land use change emissions in biofuel lifecycles. Several dynamic general/partial equilibrium models have been used to predict the implications of biofuel policy and commercialization on land use, international/domestic trade, and GHG fluxes and commodity markets within the agricultural sector including the Global Trade Analysis Project (GTAP), the Market Allocation Model, the Forestry and Agricultural Sector Optimization Model (FASOM), and the Food and Agricultural Policy Research Institute model, LEI-TAP model, and Modular Applied General Equilibrium Tool, Center for Agricultural and Rural Development model, International Model for Policy Analysis of Agricultural Commodities and Trade model, Common Agricultural Policy Regional Impact Analysis model, the Worldwide Agribusiness Linkage Program and Commodity Simulation Model, and the Modeling International Relationships in Applied General Equilibrium model [129,130,131,132,133,134]. However, lack of model transparency and high complexity often limit the utility of these approaches. Furthermore, these models often vary in regards to data requirements, scope, and model resolution. For example, models such as FASOM have high resolution for the United States but lack information regarding international trade; while models such as GTAP can provide estimates for domestic and international land-use change but at a low level resolution. Synthesis and coupling these dynamic economic models with biophysical land-use models such as the Integrated Model to Assess the Greenhouse Effect, Conversion of Land Use and its Effects as well as energy models such as the PRIMES Energy System Model can help support informed decision making [135,136]. However, it is important to note that that economic systems exhibit structural inconstancy (*i.e.*, change in individual behavior in response to a policy change), and coupling economic and biophysical models will increase overall model uncertainty [137]. Plevin *et al.* pair an economic-computable general equilibrium model with a CO_2 emissions estimation model and conclude that due to parametric uncertainties, the results obtained should be used for developing a range of possible results for comparison purposes rather than as deterministic estimates of land use change emissions [138].

LCAs of emerging technologies are difficult to conduct due to lack of technology specific data, dynamic and rapidly evolving systems, and isolation of environmental research from technical developments [139]. Prospective LCAs involve estimating environmental impacts of possible future scenarios and are affected by the choice of time horizon, complexity of the system, and extent of stakeholder engagement. Studies have used approaches such as scenario analysis [140,141], participatory methods involving expert elicitations and stakeholder opinions [139] and modeling of economic transactions to understand market-mediated effects [142]. Many of the challenges associated with prospective LCA can be overcome with inclusion of life cycle thinking at an early stage in research and development (R&D) via a collaborative dialogue between industry experts, stakeholders, and LCA practitioners.

Analysis at the supply chain or life cycle scale is often focused on improving efficiency and reducing environmental impacts across the life cycle, utilizing methods and metrics such as eco-efficiency, life cycle GHG emissions, net energy analysis, water footprint, and others. However, these methods fail to capture the impact of biofuel life cycles on ecological goods and services, *i.e.*, the fundamental goods and services provided by nature that sustain human life and are the fodder for all man-made capital and industrial activity [143]. Consequently, decisions based on these methods could result in unsustainable choices including heightened depletion and degradation of natural capital and ecosystems [111,144,145].

4.4. Ecosystems Scale

The final tier of analysis extends the analytic scope to consider the role of ecological goods and services throughout the supply chain. As ecological goods and services play a fundamental role in sustaining industrial activity, it is paramount to account for their role in evaluating the impact and sustainability of emerging biofuel platforms. Examples of ecological goods and services include: timber, food, water, energy resources, clean air, minerals and ores, purification of air and water resources, flood and drought mitigation, pollination of crops and vegetation, maintenance of global biodiversity, as well as climate and disease regulation [146,147,148]. Since it is computationally intractable to model the complete supply chain by including each process as done at the life cycle level of analysis, the approach leveraged at this scale of analysis is closely related to existing hybrid (*i.e.*, tiered) LCA methods in which a process level model is used to determine process level consumption of ecological goods and services while economy wide-impacts are incorporated using EcoLCA [108,109,112,116,149].

Process level flows of ecological goods and services can be modeled using a host of computational models. Detailed models such as Century, EPIC, APEX, and SWAT can be used to simulate the effects of land management decisions on soil, water, nutrients and watersheds; however, these tools are data intensive and suffer from high model complexity [150,151,152]. The Integrated Valuation of Ecosystems Services and Tradeoffs (InVEST) modeling suite can be used to quantify and map a variety of regional ecosystems goods services as well as biodiversity for both terrestrial and marine environments including crop pollination, habitat quality, habitat risk assessment, managed timber production, managed fish aquaculture, marine water quality, sediment retention, water purification, carbon sequestration, and others [153]. Further, InVEST model(s) are open-source, generally simpler, more transparent, and user friendly as compared to the aforementioned approaches. Synthesizing these computational models with EcoLCA can provide a holistic understanding of the potential impacts and tradeoffs of biofuel production on ecological resources and biodiversity. Although this tier of analysis is the most comprehensive it often has high uncertainty due to (i) variability that is propagated and compounded at each preceding level of analysis (*i.e.*, from lab/field scale, to process scale, to supply chain); and (ii) high uncertainty in the modeling approaches used to quantify ecological goods and services.

Bakshi and colleagues have developed and applied a hybrid EcoLCA framework to investigate the sustainability of petroleum transportation fuels as well as select first and second-generation biofuels [110,111]. More recently, this framework has been applied to study emerging microalgal biofuel systems [145]. The results of these studies reveal that biofuels have high renewability but typically have low thermodynamic return on investment relative to baseline petroleum fuels. The low energy return on investment for biofuels is concerning as prior studies have suggested that a liquid technical fuel must achieve minimum threshold energy return on investment (EROI) values to sustain society. Failure to meet this minimum EROI criterion could result in widespread economic and social ramifications as more useful work must be expended by society for fuel production and thus cannot sustain other economic activities [154,155]. Recently, Bakshi and colleagues have developed a conceptual framework for designing technological and ecological systems that encourages synergy between human activity and nature [70,156]. The proposed techno-ecological synergy (TES) framework considers the demand of ecological goods and services from technological systems at multiple spatial scales ranging from the individual process scale, to supply chains and life cycles, as well as the supply of ecological good and services from ecological systems ranging from regional, to watershed, to global. The TES framework aims to reduce overconsumption of natural capital, and promotes technological systems to operate within safe ecological boundaries at multiple analytic scales. Application of the TES framework can provide unique insights into the sustainability of emerging biofuel supply chains.

4.5. Accounting for Multiple Objectives and Scales in Designing Sustainable Biofuel Processes/Supply Chains

Environmentally conscious development of emerging biofuel pathways requires addressing alternatives that exist at each step along the life cycle with the possibility of a multitude of useful coproducts and waste streams. As such, decisions based on optimizing a single criterion (such as carbon footprint, EROI, or economic potential) can lead to the unintended consequence of trading one environmental problem for another. Many studies have focused on evaluating bioenergy potential encompassing several criteria—economic performance, environmental and social impact and have developed several tools that quantify these indicators [157]. For example, the SCORE Model developed by Krajnc and Domac uses a mix of qualitative and quantitative indicators such as contribution to forest management, impact of regional unemployment, CO_2 emissions, and percentage of self-sufficiency in electricity production, to analyze the sustainability of woody biomass production [158].

Studies often employ various optimization techniques to design an optimal biomass supply chain based on multiple criteria [159,160]. Formulating the design problem as a mathematical optimization task has been a common approach for analyzing technological systems. The design problem is generally formulated as either a mixed integer linear or non-linear optimization problem [161,162]. Research efforts have also resulted in coupling the design problem with LCA by quantifying the life cycle impacts of process alternatives. This could be accomplished using either single objective or multi-objective optimization (MOP) resulting in designs or options that represent the best compromise between the selected design criteria [163,164] Several studies have utilized MOP for the strategic design and implementation of biofuel systems. Zamboni *et al.* developed a spatially explicitly mixed integer linear program (MILP) optimization model for bioethanol production systems that simultaneously considers supply chain costs and life cycle GHG emissions [165,166]. De Meyer *et al.* developed a generalized mathematical model, OPTIMASS, that optimizes over strategic and tactical decisions, and can be used to investigate the potential effect of policy changes, emerging biofeedstocks, technological adoption/evolution, and logistics on the environmental sustainability of biofuel supply chains [167]. Čuček *et al.*coupled MOP with a regional biomass supply chain model [168]. Mele *et al.* developed a MILP optimization model to optimize economic and environmental objectives of the biofuel production chain, and applied the model to the sugarcane industry in Argentina [169]. More recently, Yue *et al.* developed a multi-objective life cycle optimization framework and applied it to study emerging hydrocarbon biofuel production [170]. MOP can be used to identify design solutions that optimize economic, environmental, and ecological dimensions of biofuel production; this set of optimal points constitutes a Pareto frontier in the design-solution space. Moreover, solutions that do not lie on the Pareto frontier are either infeasible or are sub-optimal. The Pareto frontier is particularly useful in MOP problems since by restricting attention to the set of choices that are Pareto efficient, a designer can evaluate tradeoffs within this set, rather than considering the full range of every parameter. Further, MOP can be used to identify the optimal mix of useful coproducts satisfying the selected set of life cycle environmental constraints. An alternative to mathematical programing is a heuristic approach that employs algorithms based on artificial intelligence to obtain a satisfactory local optimal solution, when a global optimal solution is not possible. Studies have employed various algorithms such as particle swarm optimization, genetic algorithms or honeybee foraging algorithms to identify a range of optimal solutions for various aspects of the biomass supply chain [171,172,173]. These approaches are expected to lead to the identification of synergies between feedstock production, processing methods, and the final mix of fuels and coproducts for the sustainable design of biorefineries [170,174,175].

4.6. Uncertainty and Variability

Emerging hydrocarbon biofuel platforms have a high degree of uncertainty [176], due to lack of commercialization, climatic variability, technological evolution, material and energy price volatility, variability in supply and demand dynamics, dynamic effects in ecosystems, and changes in biofuel incentives and legislation over time. Further, sources of modeling uncertainty generally include: (i) parameter uncertainty; (ii) technological uncertainty; (iii) random error; (iv) systematic uncertainty; (v) methodological uncertainty; (vi) parametric variability; (vii) structural uncertainty; (viii) algorithmic/interpolation uncertainty; and (ix) policy uncertainty. These sources of uncertainty introduce variability at each stage of the analysis, which are compounded and propagated with subsequent modeling scales and the use of higher complexity models. Several commons approaches are often utilized to quantify uncertainty in environmental sustainability analysis including stochastic modeling and one-at-a-time (OAT) sensitivity analysis [177,178,179]. It is important to note that the primary utility of environmental sustainability analysis is to identify potential environmental impacts or damages of emerging technologies at early stages of R&D. However, recommendations at the design/conceptual phase typically have high uncertainty, which is often only reduced after large investments and progress in R&D have been made. As such, environmental sustainability analysis of emerging technologies inherently faces a tradeoff between utility and uncertainty. Additionally, for biofuel production uncertainty in upstream processes can translate into heightened risk downstream. For example, farmers are often reluctant to grow second generation biofeedstocks due to their high fixed cost of production, long establishment period, and uncertainty regarding the demand for these crops. However, the lack of large-scale agricultural production of second generation biofeedstocks generates heightened risk for developing next generation biorefineries, thus potentially limiting the demand for second-generation biofeedstocks.

5. CONCLUSIONS AND OUTLOOK

The framework outlined in this study has multiple features that make it conceptually attractive; however, the approach faces several challenges: (I) Interdisciplinary research requires effective communication, transfer, and synthesis of domain specific knowledge and data across research fields. This requires that collaborators become proficient in the colloquial terminology commonly used in said research fields, so that domain specific modeling results and technical information can be exchanged efficiently and information is not lost at the interface of research fields. (II) The long timeframe required for laboratory/field trials as well as the development and implementation of modeling tools may limit the effectiveness of this approach, as the "state of the art" may have changed or the technical system evolved from its prior conception at the start of the analysis. However, this issue can be addressed via several means. Streamlined models and methods can help reduce the computation time necessary for model development and implementation. Further, open source databases and the use of "big data" may help reduce the time required for data acquisition. Additionally, the modeling approach should be iterative; models should be configured to use the most robust, accurate and up-to-date data, with modeling tiers providing feedback-loops to inform the strategic design and development of biofuel processes and supply chains. (III) Differing models/tools may have contrasting assumptions, conflicting theoretical premises, and may report the results using disparate metrics. Additionally, the results may be subjective and have high degree of uncertainty. For these reasons, results should not be considered deterministic outcomes, but probabilistic measures to be used for comparative purposes [180].

In its inception, first generation biofuels were touted as a revolutionary new renewable and sustainable fuel source. Technological exuberance and naivety lead to their large-scale commercialization without proper consideration of the potential consequences for industry and the environment. Since then, a multitude of studies have shown that the production and use first generation biofuels has resulted in detrimental impacts on the ecosphere, environment, economy, and social welfare [64,181]. As biofuel production is inherently interconnected with various critical sectors of the economy (*i.e.*, agriculture, transportation, *etc.*), it is crucial to understand the potential widespread impact of biofuels on economics, environment, and human welfare before their widespread adoption and commercialization. Accordingly, a new paradigm is needed for the design of sustainable biofuel processes supply chains, and is contingent on collaboration across research domains and evaluation of potential impacts over multiple spatial and temporal scales. This work discussed a novel modular multiscale and multiobjective framework for analyzing emerging biofuel supply chains. While process-scale analysis is the most reductionist and is commonly employed in engineering analysis and traditional process design, this narrowly focused approach fails to capture broader environmental externalities; such shortsightedness could jeopardize the sustainability of emerging biofuel systems. The conceptual multiscale and multiobjective framework presented in this work addresses these limitations by optimizing economic and environmental objectives over the field/lab scale, process scale, supply chain, and ecosystems scale. This broad-based interdisciplinary approach is critical for guiding the sustainable development of the biofuels industry and for mitigating and avoiding any unintended consequences.

Further, the results of the bibliographic analysis support the need for such a collaborative framework. Next generation biofuels represent a promising opportunity for the chemical industry and sustainability engineers to work hand in hand, and transform the traditional paradigm from "end of pipe" solutions to innovative, state of the art, and sustainable design solutions.

ACKNOWLEDGMENTS

This material is based upon work supported by the National Science Foundation Graduate Research Fellowship under Grant No. (DGE-1247842). Any opinion, findings, and conclusions or recommendations expressed in this material are those of the authors(s) and do not necessarily reflect the views of the National Science Foundation. This work is supported by the National Institute of Food and Agriculture, United States Department of Agriculture (USDA) (Agriculture and Food Research Initiative Competitive Grant No. 2012-67009-19717). Any opinions, findings and conclusions or recommendations expressed in this material are those of the authors and do not necessarily reflect the views of USDA.

AUTHOR CONTRIBUTIONS

GGZ and NV wrote the manuscript and conducted the bibliometric analysis. SSC performed the network analysis and constructed the bibliometric network figure. VK coordinated the study, and assisted/guided in the development and analysis of the network figure. VK and AEL reviewed the manuscript. All authors read and approved the final manuscript.

REFERENCES

1. Sarukhan, J.; Whyte, A.; Hassan, R.; Scholes, R.; Ash, N.; Carpenter, S.; Pingali, P.; Bennett, E.; Zurek, M.; Chopra, K. *Millenium ecosystem assessment: Ecosystems and human well-being*; WHO Press: Geneva, Switzerland, 2005.
2. Steffen, W.; Richardson, K.; Rockström, J.; Cornell, S.E.; Fetzer, I.; Bennett, E.M.; Biggs, R.; Carpenter, S.R.; de Vries, W.; de Wit, C.A.; *et al.* Planetary boundaries: Guiding human development on a changing planet.*Science* **2015**.

3. Hughes, T.P.; Carpenter, S.; Rockström, J.; Scheffer, M.; Walker, B. Multiscale regime shifts and planetary boundaries. *Trends Ecol. Evol.* **2013**, *28*, 389–395.
4. Steffen, W.; Persson, Å.; Deutsch, L.; Zalasiewicz, J.; Williams, M.; Richardson, K.; Crumley, C.; Crutzen, P.; Folke, C.; Gordon, L.; *et al.* The anthropocene: From global change to planetary stewardship. *AMBIO* **2011**,*40*, 739–761.
5. Rockstrom, J.; Steffen, W.; Noone, K.; Persson, A.; Chapin, F.S.; Lambin, E.F.; Lenton, T.M.; Scheffer, M.; Folke, C.; Schellnhuber, H.J.; *et al.* A safe operating space for humanity. *Nature* **2009**, *461*, 472–475.
6. EPA. *Inventory of U.S. Greenhouse Gas Emissions and Sinks: 1990–2011*; United States Environmental Protection Agency: Washington, DC, USA, 2015.
7. Sissine, F. *Energy Independence and Security Act of 2007: A Summary of Major Provisions*; CRS Order Code RL34294. Library of Congress, Congressional Research Service: Washington, DC, USA, 2007.
8. European Parliament and the Council of the European Union. Directive 2009/28/EC of the european parliament and of the council of 23 April 2009 on the promotion of the use of energy from renewable sources and amending and subsequently repealing directives 2001/77/EC and 2003/30/EC. *Off. J. Eur. Union***2009**, *52*, 16–62.
9. Fargione, J.; Hill, J.; Tilman, D.; Polasky, S.; Hawthorne, P. Land clearing and the biofuel carbon debt.*Science* **2008**, *319*, 1235–1238.
10. Melillo, J.M.; Reilly, J.M.; Kicklighter, D.W.; Gurgel, A.C.; Cronin, T.W.; Paltsev, S.; Felzer, B.S.; Wang, X.D.; Sokolov, A.P.; Schlosser, C.A. Indirect emissions from biofuels: How important? *Science* **2009**, *326*, 1397–1399.
11. McKone, T.E.; Nazaroff, W.W.; Berck, P.; Auffhammer, M.; Lipman, T.; Torn, M.S.; Masanet, E.; Lobscheid, A.; Santero, N.; Mishra, U.; *et al.* Grand challenges for life-cycle assessment of biofuels. *Environ. Sci. Technol.* **2011**, *45*, 1751–1756.
12. Williams, P.R.D.; Inman, D.; Aden, A.; Heath, G.A. Environmental and sustainability factors associated with next-generation biofuels in the U.S.: What do we really know? *Environ. Sci. Technol.* **2009**, *43*, 4763–4775.

13. Perlack, R.D.; Stokes, B.J. *Us Billion-ton Update: Biomass Supply for a Bioenergy and Bioproducts Industry*; Oak Ridge National Laboratory: Oak Ridge, TN, USA, 2011.
14. Nigam, P.S.; Singh, A. Production of liquid biofuels from renewable resources. *Prog. Energy Combust. Sci.* **2011**, *37*, 52–68.
15. Sims, R.E.H.; Mabee, W.; Saddler, J.N.; Taylor, M. An overview of second generation biofuel technologies. *Bioresour. Technol.* **2010**, *101*, 1570–1580.
16. Huber, G.W.; Dumesic, J.A. An overview of aqueous-phase catalytic processes for production of hydrogen and alkanes in a biorefinery. *Catal. Today* **2006**, *111*, 119–132.
17. Huber, G.W.; Iborra, S.; Corma, A. Synthesis of transportation fuels from biomass: Chemistry, catalysts, and engineering. *Chem. Rev.* **2006**, *106*, 4044–4098.
18. Lee, S.K.; Chou, H.; Ham, T.S.; Lee, T.S.; Keasling, J.D. Metabolic engineering of microorganisms for biofuels production: From bugs to synthetic biology to fuels. *Curr. Opin. Biotechnol.* **2008**, *19*, 556–563.
19. Chheda, J.N.; Dumesic, J.A. An overview of dehydration, aldol-condensation and hydrogenation processes for production of liquid alkanes from biomass-derived carbohydrates. *Catal. Today* **2007**, *123*, 59–70.
20. Chheda, J.N.; Huber, G.W.; Dumesic, J.A. Liquid-phase catalytic processing of biomass-derived oxygenated hydrocarbons to fuels and chemicals. *Angew. Chem. Int. Ed.* **2007**, *46*, 7164–7183.
21. Chheda, J.N.; Roman-Leshkov, Y.; Dumesic, J.A. Production of 5-hydroxymethylfurfural and furfural by dehydration of biomass-derived mono- and poly-saccharides. *Green Chem.* **2007**, *9*, 342–350.
22. Huber, G.W.; Chheda, J.N.; Barrett, C.J.; Dumesic, J.A. Production of liquid alkanes by aqueous-phase processing of biomass-derived carbohydrates. *Science* **2005**, *308*, 1446–1450.
23. Huber, G.W.; Shabaker, J.W.; Dumesic, J.A. Raney Ni-Sn catalyst for H_2 production from biomass-derived hydrocarbons. *Science* **2003**, *300*, 2075–2077.
24. Olcay, H.; Subrahmanyam, A.V.; Xing, R.; Lajoie, J.; Dumesic, J.A.; Huber, G.W. Production of renewable petroleum refinery diesel and jet fuel feedstocks from hemicellulose sugar streams. *Energy Environ. Sci.* **2013**, *6*, 205–216.

25. Bond, J.Q.; Upadhye, A.A.; Olcay, H.; Tompsett, G.A.; Jae, J.; Xing, R.; Alonso, D.M.; Wang, D.; Zhang, T.; Kumar, R.; *et al*. Production of renewable jet fuel range alkanes and commodity chemicals from integrated catalytic processing of biomass. *Energy Environ. Sci.* **2014**, *7*, 1500–1523.
26. Bond, J.Q.; Alonso, D.M.; Wang, D.; West, R.M.; Dumesic, J.A. Integrated catalytic conversion of γ-valerolactone to liquid alkenes for transportation fuels. *Science* **2010**, *327*, 1110–1114.
27. Zaimes, G.G.; Soratana, K.; Harden, C.L.; Landis, A.E.; Khanna, V. Biofuels via fast pyrolysis of perennial grasses: A life cycle evaluation of energy consumption and greenhouse gas emissions. *Environ. Sci. Technol.***2015**, *49*, 10007–10018.
28. Bridgwater, A.V. Review of fast pyrolysis of biomass and product upgrading. *Biomass Bioenergy* **2012**, *38*, 68–94.
29. French, R.; Czernik, S. Catalytic pyrolysis of biomass for biofuels production. *Fuel Process. Technol.* **2010**, *91*, 25–32.
30. Wright, M.M.; Daugaard, D.E.; Satrio, J.A.; Brown, R.C. Techno-economic analysis of biomass fast pyrolysis to transportation fuels. *Fuel* **2010**, *89*, S2–S10.
31. Patwardhan, P.R.; Brown, R.C.; Shanks, B.H. Understanding the fast pyrolysis of lignin. *ChemSusChem***2011**, *4*, 1629–1636.
32. Patwardhan, P.R.; Brown, R.C.; Shanks, B.H. Product distribution from the fast pyrolysis of hemicellulose.*ChemSusChem* **2011**, *4*, 636–643.
33. Elliott, D.C.; Hart, T.R.; Neuenschwander, G.G.; Rotness, L.J.; Zacher, A.H. Catalytic hydroprocessing of biomass fast pyrolysis bio-oil to produce hydrocarbon products. *Environ. Prog. Sustain. Energy* **2009**, *28*, 441–449.
34. Wang, K.; Brown, R.C. Catalytic pyrolysis of corn dried distillers grains with solubles to produce hydrocarbons. *ACS Sustain. Chem. Eng.* **2014**, *2*, 2142–2148.
35. Brewer, C.E.; Schmidt-Rohr, K.; Satrio, J.A.; Brown, R.C. Characterization of biochar from fast pyrolysis and gasification systems. *Environ. Prog. Sustain. Energy* **2009**, *28*, 386–396.
36. Graber, E.; Harel, Y.M.; Kolton, M.; Cytryn, E.; Silber, A.; David, D.R.; Tsechansky, L.; Borenshtein, M.; Elad, Y. Biochar impact on development and productivity of pepper and tomato grown in fertigated soilless media. *Plant Soil* **2010**, *337*, 481–496.

37. Laird, D.A.; Fleming, P.; Davis, D.D.; Horton, R.; Wang, B.; Karlen, D.L. Impact of biochar amendments on the quality of a typical midwestern agricultural soil. *Geoderma* **2010**, *158*, 443–449.
38. Major, J.; Lehmann, J.; Rondon, M.; Goodale, C. Fate of soil-applied black carbon: Downward migration, leaching and soil respiration. *Glob. Change Biol.* **2010**, *16*, 1366–1379.
39. Vaccari, F.P.; Baronti, S.; Lugato, E.; Genesio, L.; Castaldi, S.; Fornasier, F.; Miglietta, F. Biochar as a strategy to sequester carbon and increase yield in durum wheat. *Eur. J. Agron.* **2011**, *34*, 231–238.
40. Verheijen, F.G.A.; Graber, E.R.; Ameloot, N.; Bastos, A.C.; Sohi, S.; Knicker, H. Biochars in soils: New insights and emerging research needs. *Eur. J. Soil Sci.* **2014**, *65*, 22–27.
41. Woolf, D.; Amonette, J.E.; Street-Perrott, F.A.; Lehmann, J.; Joseph, S. Sustainable biochar to mitigate global climate change. *Nat. Commun.* **2010**, *1*, 56.
42. Agrawal, R.; Singh, N.R. Synergistic routes to liquid fuel for a petroleum-deprived future. *AIChE J.* **2009**,*55*, 1898–1905.
43. Dauenhauer, P.J.; Dreyer, B.J.; Degenstein, N.J.; Schmidt, L.D. Millisecond reforming of solid biomass for sustainable fuels. *Angew. Chem. Int. Ed.* **2007**, *46*, 5864–5867.
44. Brune, D.E.; Lundquist, T.J.; Benemann, J.R. Microalgal Biomass for Greenhouse Gas Reductions: Potential for Replacement of Fossil Fuels and Animal Feeds. *J. Environ. Eng.* **2009**, *135*, 1136–1144.
45. Weyer, K.M.; Bush, D.R.; Darzins, A.; Willson, B.D. Theoretical maximum algal oil production. *BioEnergy Res.* **2010**, *3*, 204–213.
46. Chisti, Y. Biodiesel from microalgae. *Biotechnol. Adv.* **2007**, *25*, 13.
47. Yang, J.; Xu, M.; Zhang, X.; Hu, Q.; Sommerfeld, M.; Chen, Y. Life-cycle analysis on biodiesel production from microalgae: Water footprint and nutrients balance. *Bioresour. Technol.* **2011**, *102*, 159–165.
48. Brennan, L.; Owende, P. Biofuels from microalgae—A review of technologies for production, processing, and extractions of biofuels and co-products. *Renew. Sustain. Energy Rev.* **2010**, *14*, 557–577.
49. Luo, D.; Hu, Z.; Choi, D.G.; Thomas, V.M.; Realff, M.J.; Chance, R.R. Life cycle energy and greenhouse gas emissions for an ethanol production process based on blue-green algae. *Environ. Sci. Technol.* **2010**, *44*, 8670–8677.

50. Ehimen, E.A.; Sun, Z.F.; Carrington, C.G.; Birch, E.J.; Eaton-Rye, J.J. Anaerobic digestion of microalgae residues resulting from the biodiesel production process. *Appl. Energy* **2011**, *88*, 3454–3463.
51. Sialve, B.; Bernet, N.; Bernard, O. Anaerobic digestion of microalgae as a necessary step to make microalgal biodiesel sustainable. *Biotechnol. Adv.* **2009**, *27*, 409–416.
52. Tilman, D.; Hill, J.; Lehman, C. Carbon-negative biofuels from low-input high-diversity grassland biomass.*Science* **2006**, *314*, 1598–1600.
53. Prochnow, A.; Heiermann, M.; Plöchl, M.; Amon, T.; Hobbs, P.J. Bioenergy from permanent grassland—A review: 2. Combustion. *Bioresour. Technol.* **2009**, *100*, 4945–4954.
54. Prochnow, A.; Heiermann, M.; Plöchl, M.; Linke, B.; Idler, C.; Amon, T.; Hobbs, P.J. Bioenergy from permanent grassland—A review: 1. Biogas. *Bioresour. Technol.* **2009**, *100*, 4931–4944.
55. Van Meerbeek, K.; Appels, L.; Dewil, R.; van Beek, J.; Bellings, L.; Liebert, K.; Muys, B.; Hermy, M. Energy potential for combustion and anaerobic digestion of biomass from low-input high-diversity systems in conservation areas. *GCB Bioenergy* **2015**, *7*, 888–898.
56. Soratana, K.; Harden, C.L.; Zaimes, G.G.; Rasutis, D.; Antaya, C.L.; Khanna, V.; Landis, A.E. The role of sustainability and life cycle thinking in U.S. Biofuels policies. *Energy Policy* **2014**, *75*, 316–326.
57. Naik, S.N.; Goud, V.V.; Rout, P.K.; Dalai, A.K. Production of first and second generation biofuels: A comprehensive review. *Renew. Sustain. Energy Rev.* **2010**, *14*, 578–597.
58. Demirbas, A. Political, economic and environmental impacts of biofuels: A review. *Appl. Energy* **2009**, *86*, S108–S117.
59. Soratana, K.; Khanna, V.; Landis, A.E. Re-envisioning the renewable fuel standard to minimize unintended consequences: A comparison of microalgal diesel with other biodiesels. *Appl. Energy* **2013**, *112*, 194–204.
60. Ballew, B.S. Elsevier's scopus® database.*J. Electron. Resour. Med. Libr.* **2009**, *6*, 245–252.
61. Ridley, C.E.; Clark, C.M.; LeDuc, S.D.; Bierwagen, B.G.; Lin, B.B.; Mehl, A.; Tobias, D.A. Biofuels: Network analysis of the literature reveals key environmental and economic unknowns. *Environ. Sci. Technol.* **2012**,*46*, 1309–1315.

62. Van Eck, N.J.; Waltman, L. Software survey: Vosviewer, a computer program for bibliometric mapping.*Scientometrics* **2010**, *84*, 523–538.
63. Carley, K.M.; DeReno, M. *Ora: Organization Risk Analyzer, Ora User's Guide*; Technical Report; School of Computer Science, Institute for Software Research, Carnegie Mellon University: Pittsburgh, PA, USA, 2006.
64. Dias de Oliveira, M.E.; Vaughan, B.E.; Rykiel, E.J. Ethanol as fuel: Energy, carbon dioxide balances, and ecological footprint. *BioScience* **2005**, *55*, 593–602.
65. Pimentel, D.; Patzek, T.; Cecil, G. Ethanol production: Energy, economic, and environmental losses. In*Reviews of Environmental Contamination and Toxicology*; Whitacre, D., Ware, G., Nigg, H., Doerge, D., Albert, L., Voogt, P., Gerba, C., Hutzinger, O., Knaak, J., Mayer, F., *et al*, Eds.; Springer: New York, NY, USA, 2007; Volume 189, pp. 25–41.
66. Guinee, J.B.; Heijungs, R.; Huppes, G.; Zamagni, A.; Masoni, P.; Buonamici, R.; Ekvall, T.; Rydberg, T. Life cycle assessment: Past, present, and future. *Environ. Sci. Technol.* **2010**, *45*, 90–96.
67. Clark, J.H. Green chemistry: Challenges and opportunities. *Green Chem.* **1999**, *1*, 1–8.
68. Bakshi, B.R.; Fiksel, J. The quest for sustainability: Challenges for process systems engineering. *AIChE J.***2003**, *49*, 1350–1358.
69. Bakshi, B.R. A thermodynamic framework for ecologically conscious process systems engineering. *Comput. Chem. Eng.* **2002**, *26*, 269–282.
70. Bakshi, B.R.; Ziv, G.; Lepech, M.D. Techno-ecological synergy: A framework for sustainable engineering.*Environ. Sci. Technol.* **2015**, *49*, 1752–1760.
71. Hanes, R.J.; Bakshi, B.R. Process to planet: A multiscale modeling framework for sustainable engineering.*AIChE J.* **2015**.
72. Hanes, R.J.; Bakshi, B.R. Sustainable process design by the process to planet framework. *AIChE J.* **2015**.
73. Zhuang, K.; Bakshi, B.R.; Herrgård, M.J. Multi-scale modeling for sustainable chemical production.*Biotechnol. J.* **2013**, *8*, 973–984.
74. Resasco, D.E.; Crossley, S.P. Implementation of concepts derived from model compound studies in the separation and conversion of bio-oil to fuel. *Catal. Today* **2014**.

75. Zhao, H.Y.; Li, D.; Bui, P.; Oyama, S.T. Hydrodeoxygenation of guaiacol as model compound for pyrolysis oil on transition metal phosphide hydroprocessing catalysts. *Appl. Catal. A: Gen.* **2011**, *391*, 305–310.
76. Pham, T.N.; Sooknoi, T.; Crossley, S.P.; Resasco, D.E. Ketonization of carboxylic acids: Mechanisms, catalysts, and implications for biomass conversion. *ACS Catal.* **2013**, *3*, 2456–2473.
77. Lapuerta, M.; Villajos, M.; Agudelo, J.R.; Boehman, A.L. Key properties and blending strategies of hydrotreated vegetable oil as biofuel for diesel engines. *Fuel Process. Technol.* **2011**, *92*, 2406–2411.
78. Darnoko, D.; Cheryan, M. Kinetics of palm oil transesterification in a batch reactor. *J. Am. Oil Chem. Soc.* **2000**, *77*, 1263–1267.
79. Nie, L.; Resasco, D.E. Kinetics and mechanism of m-cresol hydrodeoxygenation on a Pt/SiO_2 catalyst. *J. Catal.* **2014**, *317*, 22–29.
80. Christian, D.G.; Riche, A.B.; Yates, N.E. Growth, yield and mineral content of *Miscanthus giganteus* grown as a biofuel for 14 successive harvests. *Ind. Crops Prod.* **2008**, *28*, 320–327.
81. Guretzky, J.; Biermacher, J.; Cook, B.; Kering, M.; Mosali, J. Switchgrass for forage and bioenergy: Harvest and nitrogen rate effects on biomass yields and nutrient composition. *Plant Soil* **2011**, *339*, 69–81.
82. Thomason, W.E.; Raun, W.R.; Johnson, G.V.; Taliaferro, C.M.; Freeman, K.W.; Wynn, K.J.; Mullen, R.W. Switchgrass response to harvest frequency and time and rate of applied nitrogen. *J. Plant Nutr.* **2005**, *27*, 1199–1226.
83. Vicente, G.; Coteron, A.; Martinez, M.; Aracil, J. Application of the factorial design of experiments and response surface methodology to optimize biodiesel production. *Ind. Crops Prod.* **1998**, *8*, 29–35.
84. Nair, S.S.; Kang, S.; Zhang, X.; Miguez, F.E.; Izaurralde, R.C.; Post, W.M.; Dietze, M.C.; Lynd, L.R.; Wullschleger, S.D. Bioenergy crop models: Descriptions, data requirements, and future challenges. *GCB Bioenergy* **2012**, *4*, 620–633.
85. Yin, Y.; Zhang, X.; Lin, D.; Yu, H.; Wang, J.A.; Shi, P. GEPIC-VR model: A GIS-based tool for regional crop drought risk assessment. *Agric. Water Manag.* **2014**, *144*, 107–119.
86. Aspen Plus, version 11. Aspen technology Inc. Bedford, MA, USA, 2009.
87. Anex, R.P.; Aden, A.; Kazi, F.K.; Fortman, J.; Swanson, R.M.; Wright, M.M.; Satrio, J.A.; Brown, R.C.; Daugaard, D.E.; Platon, A.; *et al.*

Techno-economic comparison of biomass-to-transportation fuels via pyrolysis, gasification, and biochemical pathways. *Fuel* **2010**, *89*, S29–S35.

88. Zaimes, G.; Khanna, V. Microalgal biomass production pathways: Evaluation of life cycle environmental impacts. *Biotechnol. Biofuels* **2013**, *6*, 88.
89. Zaimes, G.G.; Khanna, V. Environmental sustainability of emerging algal biofuels: A comparative life cycle evaluation of algal biodiesel and renewable diesel. *Environ. Prog. Sustain. Energy* **2013**, *32*, 926–936.
90. Agusdinata, D.B.; Zhao, F.; Ileleji, K.; DeLaurentis, D. Life cycle assessment of potential biojet fuel production in the United States. *Environ. Sci. Technol.* **2011**, *45*, 9133–9143.
91. Pourhashem, G.; Spatari, S.; Boateng, A.A.; McAloon, A.J.; Mullen, C.A. Life cycle environmental and economic tradeoffs of using fast pyrolysis products for power generation. *Energy Fuels* **2013**, *27*, 2578–2587.
92. Dang, Q.; Yu, C.; Luo, Z. Environmental life cycle assessment of bio-fuel production via fast pyrolysis of corn stover and hydroprocessing. *Fuel* **2014**, *131*, 36–42.
93. Clarens, A.F.; Nassau, H.; Resurreccion, E.P.; White, M.A.; Colosi, L.M. Environmental impacts of algae-derived biodiesel and bioelectricity for transportation. *Environ. Sci. Technol.* **2011**, *45*, 7554–7560.
94. Vasudevan, V.; Stratton, R.W.; Pearlson, M.N.; Jersey, G.R.; Beyene, A.G.; Weissman, J.C.; Rubino, M.; Hileman, J.I. Environmental performance of algal biofuel technology options. *Environ. Sci. Technol.* **2012**,*46*, 2451–2459.
95. Wang, M.; Han, J.; Dunn, J.B.; Cai, H.; Elgowainy, A. Well-to-wheels energy use and greenhouse gas emissions of ethanol from corn, sugarcane and cellulosic biomass for us use. *Environ. Res. Lett.* **2012**, *7*, 045905.
96. Garcia, D.J.; You, F. Supply chain design and optimization: Challenges and opportunities. *Comput. Chem. Eng.* **2015**. in press.
97. Curran, M.A. Environmental life-cycle assessment. *Int. J. Life Cycle Assess.* **1996**, *1*, 179–179.
98. Ecoinvent Centre. *Ecoinvent Data v2.0*; Swiss Centre for Life Cycle Inventories: Dubendorf, Switzerland, 2007.

99. Ciroth, A. Ict for environment in life cycle applications openlca—A new open source software for life cycle assessment. *Int. J. Life Cycle Assess.* **2007**, *12*, 209–210.
100. USLCI. *Life-Cycle Inventory, v. 1.6. 0*; National Renewable Energy Laboratory: Golden, CO, USA, 2008.
101. Wang, M. *The Greenhouse Gases, Regulated Emissions, and Energy Use in Transportation Model: Version 1.5*; Argonne National Laboratory, U.S. Department of Energy: Lemont, IL, USA, 2008.
102. Zaimes, G.G.; Khanna, V. The role of allocation and coproducts in environmental evaluation of microalgal biofuels: How important? *Sustain. Energy Technol. Assess.* **2014**, *7*, 247–256.
103. Wang, M.; Huo, H.; Arora, S. Methods of dealing with co-products of biofuels in life-cycle analysis and consequent results within the U.S. Context. *Energy Policy* **2011**, *39*, 5726–5736.
104. Liu, X.; Clarens, A.F.; Colosi, L.M. Algae biodiesel has potential despite inconclusive results to date.*Bioresour. Technol.* **2012**, *104*, 803–806.
105. Sieverding, H.L.; Bailey, L.M.; Hengen, T.J.; Clay, D.E.; Stone, J.J. Meta-analysis of soybean-based biodiesel.*J. Environ. Qual.* **2015**.
106. Leontief, W. Input-output analysis. *New Palgrave. Dict. Econ.* **1987**, *2*, 860–864.
107. Lave, L.B. Using input-output analysis to estimate economy-wide discharges. *Environ. Sci. Technol.* **1995**, *29*, 420A–426A.
108. Zhang, Y.; Baral, A.; Bakshi, B.R. Accounting for ecosystem services in life cycle assessment, part ii: Toward an ecologically based LCA. *Environ. Sci. Technol.* **2010**, *44*, 2624–2631.
109. Zhang, Y.; Singh, S.; Bakshi, B.R. Accounting for ecosystem services in life cycle assessment, part i: A critical review. *Environ. Sci. Technol.* **2010**, *44*, 2232–2242.
110. Baral, A.; Bakshi, B.R. Thermodynamic metrics for aggregation of natural resources in life cycle analysis: Insight via application to some transportation fuels. *Environ. Sci. Technol.* **2009**, *44*, 800–807.
111. Baral, A.; Bakshi, B.R.; Smith, R.L. Assessing resource intensity and renewability of cellulosic ethanol technologies using eco-LCA. *Environ. Sci. Technol.* **2012**, *46*, 2436–2444.
112. Ukidwe, N.U.; Bakshi, B.R. Thermodynamic accounting of ecosystem contribution to economic sectors with application to 1992 U.S. Economy. *Environ. Sci. Technol.* **2004**, *38*, 4810–4827.

113. Sciubba, E. On the second-law inconsistency of emergy analysis. *Energy* **2010**, *35*, 3696–3706.

114. Sciubba, E.; Ulgiati, S. Emergy and exergy analyses: Complementary methods or irreducible ideological options? *Energy* **2005**, *30*, 1953–1988.

115. Dewulf, J.; van Langenhove, H.; Muys, B.; Bruers, S.; Bakshi, B.R.; Grubb, G.F.; Paulus, D.M.; Sciubba, E. Exergy: Its potential and limitations in environmental science and technology. *Environ. Sci. Technol.* **2008**,*42*, 2221–2232.

116. Hau, J.L.; Bakshi, B.R. Promise and problems of emergy analysis. *Ecol. Model.* **2004**, *178*, 215–225.

117. Gaudreau, K.; Fraser, R.; Murphy, S. The tenuous use of exergy as a measure of resource value or waste impact. *Sustainability* **2009**, *1*, 1444–1463.

118. Rugani, B.; Benetto, E. Improvements to emergy evaluations by using life cycle assessment. *Environ. Sci. Technol.* **2012**, *46*, 4701–4712.

119. Raugei, M.; Rugani, B.; Benetto, E.; Ingwersen, W.W. Integrating emergy into LCA: Potential added value and lingering obstacles. *Ecol. Model.* **2014**, *271*, 4–9.

120. Romero, J.C.; Linares, P. Exergy as a global energy sustainability indicator. A review of the state of the art.*Renew. Sustain. Energy Rev.* **2014**, *33*, 427–442.

121. Suh, S.; Lenzen, M.; Treloar, G.J.; Hondo, H.; Horvath, A.; Huppes, G.; Jolliet, O.; Klann, U.; Krewitt, W.; Moriguchi, Y.; *et al.* System boundary selection in life-cycle inventories using hybrid approaches. *Environ. Sci. Technol.* **2004**, *38*, 657–664.

122. Ekvall, T.; Weidema, B. System boundaries and input data in consequential life cycle inventory analysis.*Int. J. Life Cycle Assess.* **2004**, *9*, 161–171.

123. Schmidt, J. System delimitation in agricultural consequential LCA. *Int. J. Life Cycle Assess.* **2008**, *13*, 350–364.

124. Weidema, B.; Frees, N.; Nielsen, A.-M. Marginal production technologies for life cycle inventories. *Int. J. Life Cycle Assess.* **1999**, *4*, 48–56.

125. Earles, J.M.; Halog, A. Consequential life cycle assessment: A review. *Int. J. Life Cycle Assess.* **2011**, *16*, 445–453.

126. Reinhard, J.; Zah, R. Global environmental consequences of increased biodiesel consumption in Switzerland: Consequential life cycle assessment. *J. Clean. Prod.* **2009**, *17*, S46–S56.
127. Lund, H.; Mathiesen, B.; Christensen, P.; Schmidt, J. Energy system analysis of marginal electricity supply in consequential LCA. *Int. J. Life Cycle Assess.* **2010**, *15*, 260–271.
128. Lesage, P.; Deschênes, L.; Samson, R. Evaluating holistic environmental consequences of brownfield management options using consequential life cycle assessment for different perspectives. *Environ. Manag.* **2007**, *40*, 323–337.
129. Britz, W. *Capri Modelling System Documentation (Common Agricultural Policy Regional Impact Analysis)*; Institute for Agricultural Policy, Market Research and Economic Sociology, University Bonn: Bonn, Germany, 2005.
130. Westhoff, P.; Baur, R.; Stephens, D.L.; Meyers, W.H. *Fapri Us Crops Model Documentation*; Center for Agricultural and Rural Development, Iowa State University: Ames, IA, USA, 1990.
131. Rosegrant, M.W.; Msangi, S.; Ringler, C.; Sulser, T.B.; Zhu, T.; Cline, S.A. *International Model for Policy Analysis Agricultural Commodities and Trade (Impact): Model Description*; International Food Policy Research Institute: Washington, DC, USA, 2008.
132. Adams, D.M.; Alig, R.J.; Callaway, J.M.; McCarl, B.A.; Winnett, S.M. *The Forest and Agricultural Sector Optimization Model (Fasom): Model Structure and Policy Applications*; U.S. Department of Agriculture, Pacific Northwest Research Station: Portland, OR, USA, 1996.
133. Narayanan, B.G.; Walmsley, T.L. Global trade, assistance, and production: The GTAP 7 data base. Center for Global Trade Analysis, Purdue University. Available online: https://www.gtap.agecon.purdue.edu/databases/v7/v7_doco.asp (accessed on 13 August 2015).
134. Fishbone, L.G.; Abilock, H. Markal, a linear-programming model for energy systems analysis: Technical description of the BNL version. *Int. J. Energy Res.* **1981**, *5*, 353–375.
135. Alcamo, J. *Image 2.0: Integrated Modeling of Global Climate Change*; Kluwer Academic Publishers: Dordrecht, The Netherlands, 1994.

136. Veldkamp, A.; Fresco, L.O. Clue: A conceptual model to study the conversion of land use and its effects.*Ecol. Model.* **1996**, *85*, 253–270.
137. Scher, I.; Koomey, J. Is accurate forecasting of economic systems possible? *Clim. Change* **2011**, *104*, 473–479.
138. Plevin, R.J.; Beckman, J.; Golub, A.A.; Witcover, J.; O'Hare, M. Carbon accounting and economic model uncertainty of emissions from biofuels-induced land use change. *Environ. Sci. Technol.* **2015**, *49*, 2656–2664.
139. Wender, B.A.; Foley, R.W.; Prado-Lopez, V.; Ravikumar, D.; Eisenberg, D.A.; Hottle, T.A.; Sadowski, J.; Flanagan, W.P.; Fisher, A.; Laurin, L.; *et al.* Illustrating anticipatory life cycle assessment for emerging photovoltaic technologies. *Environ. Sci. Technol.* **2014**, *48*, 10531–10538.
140. Scown, C.D.; Nazaroff, W.W.; Mishra, U.; Strogen, B.; Lobscheid, A.B.; Masanet, E.; Santero, N.J.; Horvath, A.; McKone, T.E. Lifecycle greenhouse gas implications of us national scenarios for cellulosic ethanol production. *Environ. Res. Lett.* **2012**, *7*, 014011.
141. Brentner, L.B.; Eckelman, M.J.; Zimmerman, J.B. Combinatorial life cycle assessment to inform process design of industrial production of algal biodiesel. *Environ. Sci. Technol.* **2011**, *45*, 7060–7067.
142. Weidema, B.P. *Market Information in Life Cycle Assessment*; Miljøstyrelsen: København, Denmark, 2003; Volume 863.
143. Daily, G.C.; Polasky, S.; Goldstein, J.; Kareiva, P.M.; Mooney, H.A.; Pejchar, L.; Ricketts, T.H.; Salzman, J.; Shallenberger, R. Ecosystem services in decision making: Time to deliver. *Front. Ecol. Environ.* **2009**, *7*, 21–28.
144. Neupane, B.; Halog, A.; Lilieholm, R.J. Environmental sustainability of wood-derived ethanol: A life cycle evaluation of resource intensity and emissions in maine, USA. *J. Clean. Prod.* **2013**, *44*, 77–84.
145. Zaimes, G.G.; Khanna, V. Assessing the critical role of ecological goods and services in microalgal biofuel life cycles. *RSC Adv.* **2014**, *4*, 44980–44990.
146. Costanza, R.; d'Arge, R.; de Groot, R.; Farber, S.; Grasso, M.; Hannon, B.; Limburg, K.; Naeem, S.; O'Neill, R.V.; Paruelo, J.; *et al.* The value of the world's ecosystem services and natural capital. *Nature* **1997**, *387*, 253–260.

147. Costanza, R.; d'Arge, R.; de Groot, R.; Farber, S.; Grasso, M.; Hannon, B.; Limburg, K.; Naeem, S.; O'Neill, R.V.; Paruelo, J.; *et al.* The value of ecosystem services: Putting the issues in perspective. *Ecol. Econ.* **1998**,*25*, 67–72.

148. Farber, S.C.; Costanza, R.; Wilson, M.A. Economic and ecological concepts for valuing ecosystem services.*Ecol. Econ.* **2002**, *41*, 375–392.

149. Hau, J.L.; Bakshi, B.R. Expanding exergy analysis to account for ecosystem products and services. *Environ. Sci. Technol.* **2004**, *38*, 3768–3777.

150. Izaurralde, R.C.; McGill, W.B.; Williams, J.R. *Development and Application of the Epic Model for Carbon Cycle, Greenhouse-Gas Mitigation, and Biofuel Studies*; Pacific Northwest National Laboratory (PNNL): Richland, WA, USA, 2012.

151. Gassman, P.W.; Reyes, M.R.; Green, C.H.; Arnold, J.G. Soil and water assessment tool: Historical development, applications, and future research directions. *Trans. ASABE* **2007**.

152. Wang, X.; Gassman, P.W.; Williams, J.R.; Potter, S.; Kemanian, A.R. Modeling the impacts of soil management practices on runoff, sediment yield, maize productivity, and soil organic carbon using apex.*Soil Tillage Res.* **2008**, *101*, 78–88.

153. Tallis, H.T.; Ricketts, T.; Guerry, A.D.; Wood, S.A.; Sharp, R.; Nelson, E.; Ennaanay, D.; Wolny, S.; Olwero, N.; Vigerstol, K. InVEST 2.2.0 User's Guide. Available online: http://data.naturalcapitalproject.org/invest-releases/documentation/2_2_0/ (accessed on 13 August 2015).

154. Hall, C.A.; Balogh, S.; Murphy, D.J. What is the minimum EROI that a sustainable society must have?*Energies* **2009**, *2*, 25–47.

155. Hall, C.A.S.; Lambert, J.G.; Balogh, S.B. EROI of different fuels and the implications for society. *Energy Policy* **2014**, *64*, 141–152

156. Bakshi, B.R. Methods and tools for sustainable process design. *Curr. Opin. Chem. Eng.* **2014**, *6*, 69–74.

157. Cambero, C.; Sowlati, T. Assessment and optimization of forest biomass supply chains from economic, social and environmental perspectives—A review of literature. *Renew. Sustain. Energy Rev.* **2014**, *36*, 62–73.

158. Krajnc, N.; Domac, J. How to model different socio-economic and environmental aspects of biomass utilisation: Case study in selected regions in slovenia and croatia. *Energy Policy* **2007**, *35*, 6010–6020.
159. De Meyer, A.; Cattrysse, D.; Rasinmäki, J.; van Orshoven, J. Methods to optimise the design and management of biomass-for-bioenergy supply chains: A review. *Renew. Sustain. Energy Rev.* **2014**, *31*, 657–670.
160. Baños, R.; Manzano-Agugliaro, F.; Montoya, F.G.; Gil, C.; Alcayde, A.; Gómez, J. Optimization methods applied to renewable and sustainable energy: A review. *Renew. Sustain. Energy Rev.* **2011**, *15*, 1753–1766.
161. Biegler, L.T.; Grossmann, I.E.; Westerberg, A.W. *Systematic Methods for Chemical Process Design*; Prentice Hall PTR: Upper Saddle River, NJ, USA, 1997.
162. Biegler, L.T. *Nonlinear Programming: Concepts, Algorithms, and Applications to Chemical Processes*; SIAM: Philadelphia, PA, USA, 2010; Volume 10.
163. Kim, K.-J.; Smith, R.L. Systematic procedure for designing processes with multiple environmental objectives. *Environ. Sci. Technol.* **2005**, *39*, 2394–2405.
164. Hoffmann, V.H.; Hungerbühler, K.; McRae, G.J. Multiobjective screening and evaluation of chemical process technologies. *Ind. Eng. Chem. Res.* **2001**, *40*, 4513–4524.
165. Zamboni, A.; Bezzo, F.; Shah, N. Spatially explicit static model for the strategic design of future bioethanol production systems. 2. Multi-objective environmental optimization. *Energy Fuels* **2009**, *23*, 5134–5143.
166. Zamboni, A.; Shah, N.; Bezzo, F. Spatially explicit static model for the strategic design of future bioethanol production systems. 1. Cost minimization. *Energy Fuels* **2009**, *23*, 5121–5133.
167. De Meyer, A.; Cattrysse, D.; van Orshoven, J. A generic mathematical model to optimise strategic and tactical decisions in biomass-based supply chains (optimass). *Eur. J. Oper. Res.* **2015**, *245*, 247–264.
168. Čuček, L.; Varbanov, P.S.; Klemeš, J.J.; Kravanja, Z. Total footprints-based multi-criteria optimisation of regional biomass energy supply chains. *Energy* **2012**, *44*, 135–145.

169. Mele, F.D.; Kostin, A.M.; Guillén-Gosálbez, G.; Jiménez, L. Multiobjective model for more sustainable fuel supply chains. A case study of the sugar cane industry in argentina. *Ind. Eng. Chem. Res.* **2011**, *50*, 4939–4958.
170. Yue, D.; Kim, M.A.; You, F. Design of sustainable product systems and supply chains with life cycle optimization based on functional unit: General modeling framework, mixed-integer nonlinear programming algorithms and case study on hydrocarbon biofuels. *ACS Sustain. Chem. Eng.* **2013**, *1*, 1003–1014.
171. López, P.R.; Galán, S.G.; Reyes, N.R.; Jurado, F. A method for particle swarm optimization and its application in location of biomass power plants. *Int. J. Green Energy* **2008**, *5*, 199–211.
172. Venema, H.; Calamai, P. Bioenergy systems planning using location-allocation and landscape ecology design principles. *Ann. Oper. Res.* **2003**, *123*, 241–264.
173. Ayoub, N.; Martins, R.; Wang, K.; Seki, H.; Naka, Y. Two levels decision system for efficient planning and implementation of bioenergy production. *Energy Convers. Manag.* **2007**, *48*, 709–723.
174. Gebreslassie, B.H.; Yao, Y.; You, F. Design under uncertainty of hydrocarbon biorefinery supply chains: Multiobjective stochastic programming models, decomposition algorithm, and a comparison between cvar and downside risk. *AIChE J.* **2012**, *58*, 2155–2179.
175. You, F.; Tao, L.; Graziano, D.J.; Snyder, S.W. Optimal design of sustainable cellulosic biofuel supply chains: Multiobjective optimization coupled with life cycle assessment and input-output analysis. *AIChE J.* **2012**,*58*, 1157–1180.
176. Awudu, I.; Zhang, J. Uncertainties and sustainability concepts in biofuel supply chain management: A review. *Renew. Sustain. Energy Rev.* **2012**, *16*, 1359–1368.
177. Lloyd, S.M.; Ries, R. Characterizing, propagating, and analyzing uncertainty in life-cycle assessment: A survey of quantitative approaches. *J. Ind. Ecol.* **2007**, *11*, 161–179.
178. Lo, S.-C.; Ma, H.-W.; Lo, S.-L. Quantifying and reducing uncertainty in life cycle assessment using the bayesian monte carlo method. *Sci. Total Environ.* **2005**, *340*, 23–33.
179. Huijbregts, M. Uncertainty and variability in environmental life-cycle assessment. *Int. J. Life Cycle Assess.* **2002**, *7*, 173.

180. Cherubini, F.; Strømman, A.H. Life cycle assessment of bioenergy systems: State of the art and future challenges. *Bioresour. Technol.* **2011**, *102*, 437–451.

181. Yang, Y.; Bae, J.; Kim, J.; Suh, S. Replacing gasoline with corn ethanol results in significant environmental problem-shifting. *Environ. Sci. Technol.* **2012**, *46*, 3671–3678

7
CHAPTER

SCREENING AND EVALUATION OF SOME GREEN ALGAL STRAINS (CHLOROPHYCEAE) ISOLATED FROM FRESHWATER AND SODA LAKES FOR BIOFUEL PRODUCTION

Ramganesh Selvarajan[1,2,3,*], Tamás Felföldi[2], Tamás Tauber[2], Elumalai Sanniyasi[3], Timothy Sibanda[1] and Memory Tekere[1]

[1] *Department of Environmental Sciences, University of South Africa, Florida 1710, South Africa*

[2] *Department of Microbiology, Eötvös Loránd University, Pázmány Péter stny. 1/c., H-1117 Budapest, Hungary*

[3] *Department of Plant Biology and Plant Biotechnology, Presidency College, University of Madras, Chennai 600005, Tamil Nadu, India*

ABSTRACT

Microalgae are photosynthetic microorganisms that can produce lipids, proteins and carbohydrates in large amounts and within short periods of time and these can be processed into both biofuels and other useful commercial products. Due to this reason microalgae are considered as a potential source of renewable energy; and one of the most important decisions in obtaining oil from microalgae is the choice of species. In this study, the potential of Chlorophyceae species isolated from freshwater and soda lakes in Hungary and Romania (Central Europe) were characterized and evaluated by determining their biomass accumulation, lipid productivity, fatty acid profiles, and biodiesel properties besides protein and carbohydrate productivity. Out of nine strains tested, three accumulated more than 40% dry weight of protein, four accumulated more than 30% dry weight of carbohydrate and the strain Chlorella vulgaris LC8 accumulated high lipid content (42.1% ± 2.6%) with a favorable C16-C18 fatty acid profile (77.4%) as well as suitable biodiesel properties of high cetane number (57.3), low viscosity (4.7 mm2/s), lower iodine number (75.18 g I2/100 g), relative cloud point (8.8 °C) and negative cold filter plugging point (−6.5 °C). Hence the new strain, Chlorella vulgaris LC8 has potential as a feedstock for the production of excellent quality biodiesel.

Keywords: microalgae; biofuel; fresh water; soda lake; fatty acids

1. INTRODUCTION

Global consumption of crude oil is predicted to grow continuously. This explains why despite improvements in the recovery of traditional fossil fuels, more attention needs to be paid to the search for clean and viable alternative renewable energy resources with the prospect of minimizing increases in atmospheric CO_2 by recycling carbon from the atmosphere [1,2,3].

Algae are a large and highly diverse group of organisms which can be found in almost all ecosystems [4]. Microalgae are a promising feedstock for biofuel production [5,6] and different applications, such as wastewater purification [7,8,9,10], biogas production [11,12], and extraction of value added compounds for food and pharmaceutical products [13]. An important aspect of biodiesel production is the selection of a suitable algal species [14]. Selected strains should have two important key characteristics: high biomass productivity as well as adaptation to regional climatic conditions [15,16].

Many microalgae have the ability to produce amounts of triacylglycerols (TAG) of up to 50% dry cell weight as a storage lipid under photo-oxidative stress or other adverse environmental conditions [17]. Recent studies [18,19,20] prove that algae have inherent advantageous qualities like rapid biomass formation, high lipid content, tolerance for extreme environments, and thus they have generated significantly increased interest as potential feedstocks for biodiesel.

One of the potential advantages in the development and utilization of algae-derived biofuels is the greater production efficiency of the required oils compared to other fuel crops [1]. A suitable microalgae candidate for biodiesel production requires not only high lipid productivity, but also suitable fatty acid (FA) composition, as this composition can significantly influence biodiesel fuel properties such as kinematic viscosity, specific gravity, cetane number (CN), cloud point, iodine value (IV), long-chain saturated factor (LCFF) and Cold Filter Plugging Point (CFPP) [21,22]. However, inadequacies in research and development, policies and strategies for all the stages of biofuel production chain are still a limiting factor to the full exploitation of algal bioresources [14].

The isolation and characterization of oleaginous microalgae from freshwater and soda lakes from Hungary and Romania for biodiesel production is described here and this represents the first description of isolation and characterization of microalgal strains from these habitats for biofuel production.

2. RESULTS AND DISCUSSION

2.1. Sampling, Isolation and Morphological Identification of Microalgae

The sampling sites sampled in this study were all slightly alkaline, with Lake Velence having the highest pH at 9.1. Since sampling was performed in winter in this case, the Lacul Ciucas site had the lowest temperature (4.1 °C), while the others had water temperatures between 12 and 16 °C, as listed in Table 1. Microalgae represent a large variety of species living under a wide range of environmental conditions, including freshwater, lacustrine, brackish, marine and hyper-saline sites [23,24]. Previous investigations on oleaginous microalgae from different sites showed that the sampling environment plays a pivotal role in the determination of strain selection as well as strain viability [17,25,26]. For example, site specific factors, such as salinity and temperature have roles in determining lipid accumulation [27]. Microscopic observation of the nine new algal isolates confirmed their purity and allowed preliminary identification of isolates, which resulted in their classification into the class*Chlorophyceae* based on the morphological characters represented in Table 1.

2.2. Molecular Identification and Phylogenetic Analysis

The 18S rRNA gene is commonly used for the molecular identification of microeukaryotes [26,28]. Polymerase chain reaction (PCR) amplification of the genomic DNA of the algal isolates with the selected primers revealed efficient amplification. The PCR products of small sub unit (SSU) rRNA with a size of ~980 bp were recorded for all isolates. Sequencing results revealed that all isolated strains are chlorophytes *(Chlorellaceae, Chlorophyta)* belonging to three different genera: *Chlorella, Dictyosphaerium* and *Micractinium* (Table 2).

Phylogenetic analysis of the 18S rRNA gene sequences (Figure 1) affiliated the isolates LF5, LC2, LC8 and LC9 (Lake Feneketlen (LF); Lacul Ciucas (LC)) clearly to *Chlorella vulgaris*, with the closest similarity to microalgal strain *C. vulgaris* SAG 211-11b. Strains RP1 and IL2 (Lake Velence-Reed Pond (RP); Inner Lake (IL) Tihany) were identified as *Chlorella sorokiniana* and *Dictyosphaerium ehrenbergianum*, with the closest similarities to *C. sorokiniana*CCAP 211/8K and *D. ehrenbergianum* CCAP 222/10, respectively. The other sequences from isolates IL3, LC3, and LC11 were related to *Micractinium* strains having pairwise nucleotide sequence similarities ranging between 99.89% and 99.66%. Members of the division *Chlorophyta* are considered to be promising for biofuel production, due to their high growth rates and ease of cultivation, however different enrichment and gating procedures are required in order to isolate strains from other taxa [28,29].

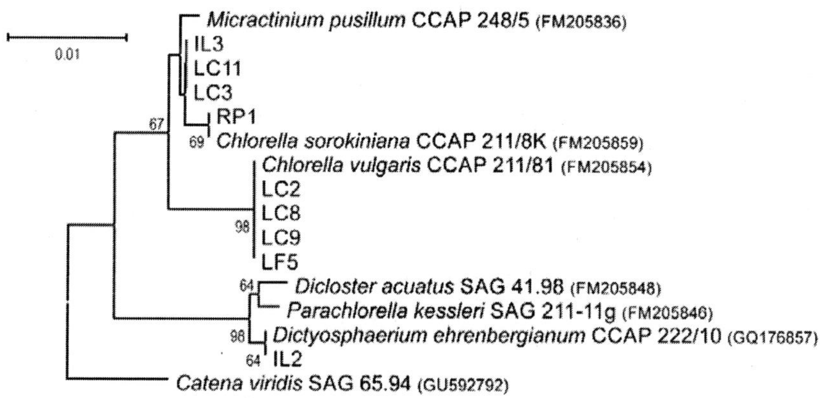

FIGURE 1. Phylogenic tree showing the relationships among the screened microalgal stains and the most similar sequences based on the 18S rRNA gene (neighbor joining method, Kimura 2-parameter nucleotide substitution model, 1000 rounds of bootstrap resampling).

TABLE 1. Sampling sites description and morphological characterization of isolated *Chlorophyceae* strains.

Genus	Sampling Site Characteristics	Strain No.	Species	Morphological Characters	Cell Dimensions	Microscopic Image
	Lake Feneketlen (LF) 47°28'36" N; 19°2'29" E, Temperature: 12.3 °C; pH: 8.6; Electrical Conductivity: 325 µS/cm	LF5	*Chlorella vulgaris*	Cells unicellular, circular, pyrenoids present, cup shaped chloroplast present, mucilagenous sheath absent	4.7 µm (length) × 4.2 µm (width)	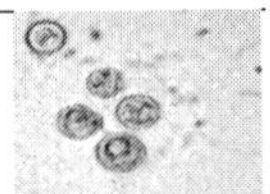
	Lacul Ciucas (LC) 46°08'44.7" N; 25°51'25.0" E; Temperature: 4.1 °C; pH: 8.1; Electrical Conductivity: 804 µS/cm	LC2	*Chlorella vulgaris*	Unicellular, circular shaped, pyrenoids present, cup shaped chloroplast, mucilagenous sheath absent	3.4 µm (length) × 3.0 µm (width)	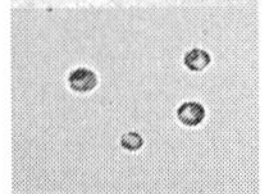
Chlorella	Lacul Ciucas (LC) see above	LC8	*Chlorella vulgaris*	Unicellular cells, cup- to girdle shaped chloroplast seen in some cells, pyrenoids present	4.0 µm (length) × 4.2 µm (width)	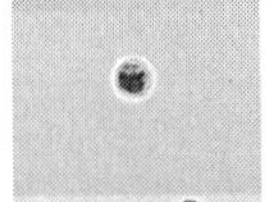
	Lacul Ciucas (LC) see above	LC9	*Chlorella vulgaris*	Cells unicellular, circular, pyrenoids present, cup shaped chloroplast present, no mucilagenous sheath observed	5.4 µm (length) × 4.9 µm (width)	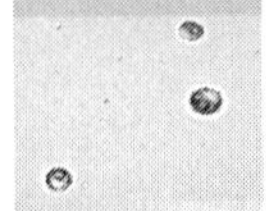
	Lake Velence-Reed Pond (RP) 47°12'40.6" N; 18°34'10.5" E; Temperature: 15.0 °C; pH: 9.1; Electrical Conductivity: 1845 µS/cm	RP1	*Chlorella sorokiniana*	Cells unicellular, circular, pyrenoids present, cup shaped chloroplast present	5.6 µm (length) × 6.1 µm (width)	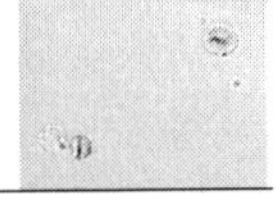
Dictyosphaerium	Inner Lake, (IL) Tihany 46°54'33.9" N; 17°53'09.7" E; Temperature: 16.1 °C; pH: 8.4; Electrical Conductivity: 860 µS/cm	IL2	*Dictyosphaerium ehrenbergianum*	Unicellular cells perpendicular to colony surface, attached to the ends of thin stalks emerging from centre of colony and branching tetrachotomously, spines are absent	4.9 µm	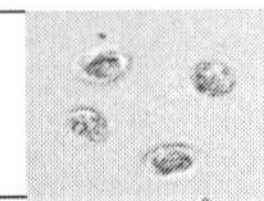
Micractinium	Inner Lake (IL) Tihany see above	IL3	*Micractinium* sp.	Solitary cells without mucilagnous sheath, planktonic, spherical, Chloroplast single parietal, cup shaped with an ellipsoidal pyrenoid	6.5 µm (length) × 5.8 µm (width)	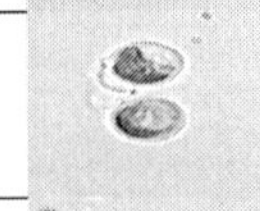
	Lacul Ciucas (LC) see above	LC3	*Micractinium* sp.	Cells are solitary without mucilaginous sheath, spherical, pyrenoid present within the chloroplast	3.9 µm (length) × 4.0 µm (width)	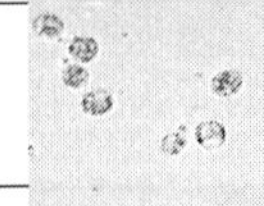
	Lacul Ciucas (LC) see above	LC11	*Micractinium* sp.	Cells are solitary without mucilaginous sheath, spherical, cup-shaped chloroplast with an ellipsoidal pyrenoid	6.0 µm (length) × 6.2 µm (width)	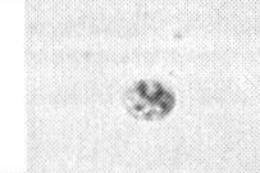

TABLE 2. Taxonomic identification of screened microalgal strains based on partial 18S rRNA gene sequence analysis.

Strain Code	Sequence Length (nt)	Closest Relative	Similarity (%)	Accession Number
LF5	925	*Chlorella vulgaris SAG 211-11b*	100	KF569724
LC2	926	*Chlorella vulgaris SAG 211-11b*	100	KF569728
LC8	907	*Chlorella vulgaris SAG 211-11b*	100	KF569734
LC9	924	*Chlorella vulgaris SAG 211-11b*	100	KF569735
RP1	919	*Chlorella sorokiniana CCAP 211/8K*	100	KF569750
IL2	926	*Dictyosphaerium ehrenbergianum CCAP 222/10*	100	KF569743
IL3	924	*Micractinium* sp.	99.89	KF569744
LC3	925	*Micractinium pusillum CCAP 248/3*	99.78	KF569729
LC11	879	*Micractinium pusillum CCAP 248/3 (and 248/1)*	99.66	KF569742

2.3. Nile Red Staining Observations

Neutral lipids including hydrocarbons and triglycerides were stained in yellow, while polar lipids were stained in red [30]. In our observations, a large number of algal cells emitted yellow fluorescence indicating the presence of neutral lipids including hydrocarbons and triglycerides as shown in Figure 2 [31,32]. The intensity of the Nile red fluorescence confirmed the presence of a large amount of lipids accumulated in the microalgal cells. The use of glycerol increases fluorescence intensity without inhibiting cell growth, thereby allowing stained cells to be isolated and cultured [33].

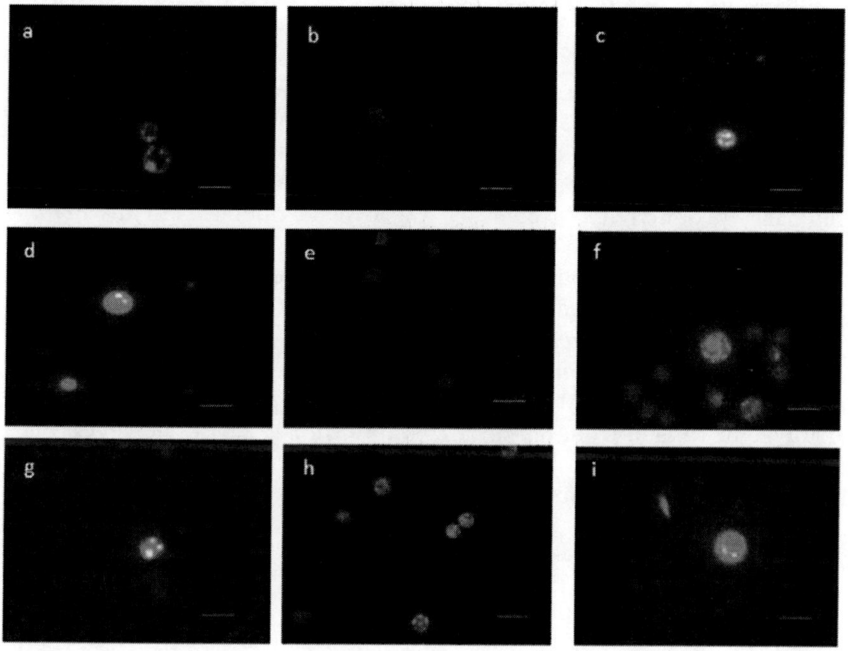

FIGURE 2. Microscopic photographs of Nile red-stained screened microalgae: (**a**) LF5; (**b**) LC2; (**c**) LC8; (**d**) LC9; (**e**) RP1; (**f**) IL2; (**g**) IL3; (**h**) LC3; (**i**) LC11.

2.4. Biomass and Lipid Content of Screened Microalgae Strains

The biomass and lipid content results are presented in Table 3. In the present study *Chlorella vulgaris* LC2 was found to be the best biomass producer (615.8 ± 10.5 mg/L). In the species *Micractinum* sp. LC3, the lowest biomass productivity was measured (294.3 ± 13.4 mg/L) as compared to the other strains where it was higher, varying between 396.0 ± 24.1 and 571.5 ± 56.0 mg/L.

Many microalgal strains naturally have high lipid contents (20%–50% dry weight). Lipid accumulation refers to increased concentration of lipids within the algal cells without consideration of the overall biomass production [34,35]. In this work however the best biomass producers did not correspond to the best lipid producers. Results shown in Table 3 reveal the notable ability of the microalgae to accumulate lipids, surpassing the total average lipid content (% of dwt) of 20% previously reported for *Chlorella* species [25,36,37,38] under the same conditions. The highest amount of lipid content was observed in *Chlorella vulgaris* LC8 (42.1% ± 2.6% dwt). In contrast, the best biomass producer, *Chlorella vulgaris* LC2, showed the lowest lipid level (8.9% ± 1.2% dwt) while in the other strains this parameter varied between 10.1 ± 1.0 and 34.1% ± 2.5% dwt. It must be noted that this study was carried out in a CO_2-limited environment which could have led to reduced biomass productivity. However, there is evidence of an inverse relationship in biomass and lipid productivity for microalgae [39]. An assessment of lipid production as a dry weight percentage of the whole biomass by the diatoms *Navicula pelliculosa*, *Navicula saprophila* and *Phaeodactylum tricornutum* [37] showed that they all produced 28%, 26% and 22% lipid dwt, respectively, which compares well with the microalgal strains in this study whose lipid production by dry weight ranged from 8.9% ± 1.2% to 42.1% ± 2.6 %. However, for a credible basis of comparison to be achieved, cultivation and production may need to be done under the same conditions.

TABLE 3. Biomass composition of screened microalgal strains.

Strain Code	Cell Dry wt (mg/L)	Protein (% dwt)	Carbohydrate (% dwt)	Lipid (% dwt)
LF5	443.8 ± 15.5	40.2 ± 8.2	19.3 ± 1.2	16.5 ± 0.1
LC2	615.8 ± 10.5	42.6 ± 5.8	24.4 ± 0.9	8.9 ± 1.2
LC8	396.0 ± 24.1	28.7 ± 1.3	20.1 ± 1.5	42.1 ± 2.6
LC9	549.7 ± 17.2	22.3 ± 0.6	43.6 ± 2.3	14.2 ± 0.1
RP1	571.5 ± 56.0	43.6 ± 3.5	10.0 ± 0.4	10.1 ± 1.0
IL2	403.3 ± 49.7	17.3 ± 2.5	40.3 ± 6.1	34.1 ± 2.5
IL3	414.3 ± 49.7	14.3 ± 3.5	41.5 ± 8.1	28.1 ± 2.5
LC3	294.3 ± 13.4	27.5 ± 2.8	38.2 ± 2.3	23.6 ± 1.8
LC11	428.1 ± 1.0	21.3 ± 1.5	13.9 ± 0.9	32.3 ± 6.7

2.5. Carbohydrate Composition

In this study, it became evident that the strains *C. vulgaris* LC9, *D. ehrenbergianum* IL2 and *Micractinum* sp. IL3 of our collection appeared to accumulate carbohydrates at up to 40% of their dry biomass (Table 3) under the conditions used. Some species such as *Chlorella*, *Dunaliella*, *Chlamydomonas* and *Scenedesmus* have been reported to accumulate more than 50% carbohydrate based on their dry cell weight; therefore microalgae are considered a promising feedstock for bioethanol production because they accumulate starch as the main carbohydrate source in their cellulose-based cell walls. Both starch and cell wall polysaccharides can be converted into fermentable sugars for subsequent bioethanol production via microbial fermentation [40]. According to the recently proposed bioconversion process of solid by-products with oleaginous microorganisms [41], they could be recycled into the lipid production process, otherwise these strains might be suitable feedstocks for bioethanol or biogas production [34]. Biobutanol can also be produced from carbohydrate-based microalgae as this alternative fuel contains more energy and is less corrosive and water soluble [42].

2.6. Protein Composition

In this work, strains LF5, LC2 and RP1 accumulated proteins up to 43% of their dry biomass under the given conditions (Table 3). Algae are natural food sources of many important aquaculture species such as molluscs, shrimps and fish [34]. According to other previous data, production of *Chlorella*, which has nutritional value and high protein content, is 2000 tonnes per annum [13]. *Chlorella* strains are also used for medicinal purposes, since they have role in improved immune response, protection against renal failure, improved fertility, better weight control, healthier skin and growth promotion of intestinal lactobacillus [34,43].

2.7. Fatty Acid Profiles

Ideal microalgal candidates for biodiesel production require not only high lipid and TAG production, but also suitable fatty acid composition [44]. Fatty Acid Methyl Esters (FAME) profiles of algal strains are given in Table 4, while comparison of the lipids with respect to the saturated, monounsaturated and polyunsaturated compounds is provided in Figure 3, which indicate that these compounds varied significantly among the nine algal strains. For example, saturated fatty acids (SFAs) ranged from 17.7% to 57.9%, monounsaturated fatty acids (MUFAs) from 1.9% to 60.6%, polyunsaturated fatty acids (PUFAs) from 4.9% to 33.8%. It is suggested that quality biodiesel should contain relatively low concentrations of both long chain saturated FAME and polyunsaturated FAME for satisfactory low temperature operability and oxidative stability [22, 45].

Palmitic acid (C16:0) was the predominant fatty acid present in the algal lipid extracts. The highest percentage was obtained with *C. vulgaris* LC2, *C. sorokiniana* RP1 and *Micractinium* sp. LC3, LC11. A reasonable balance for fuel could be achieved with oil containing high levels of monounsaturated fatty acids like palmitoleic acid (16:1) and oleic acid (18:1) because of their capability of giving the finest compromise between ignition quality, combustion heat, cold filter plugging point (CFPP), oxidative stability, viscosity, and lubricity, which are determined by the structure of its component fatty acids [22,45,46,47]. It was interesting to find out that the monounsaturated FAs composed of palmitoleic acid (C16:1) and oleic acids (C18:1) in *C. vulgaris*. LF5, LC8, LC9 and *Dictyosphaerium ehrenbergianum* IL2 presented a major percentage of about 42.4% to 60.5%, whereas other species showed small differences ranging from 0.5% to 6.6%.

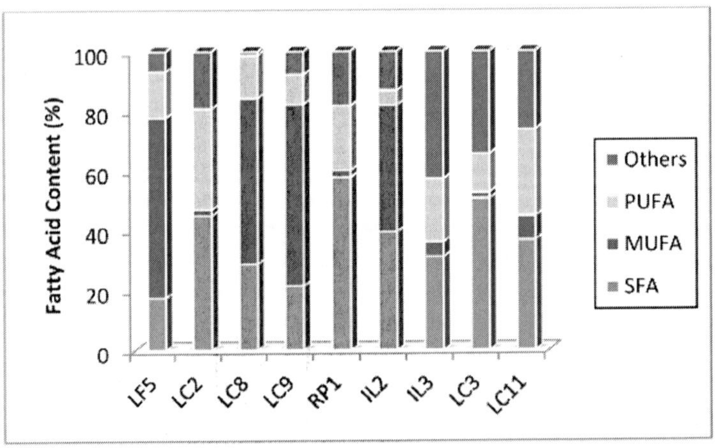

FIGURE 3. The fatty acid compositions of the screened microalgae: SFA—Saturated fatty acids (C9:0–C20:0); MUFA—Monounsaturated fatty acids (C14:1–C20:1); PUFA—Polyunsaturated fatty acids (C18:2, C20:2, C18:3, C18:3, C20:4).

TABLE 4. Fatty acid composition of the screened microalgal strains (% of total Fatty Acid Methyl Esters (FAME)).

Fatty Acids	LF5	LC2	LC8	LC9	RP1	IL2	IL3	LC3	LC11
C9:0	nd	nd	nd	nd	0.03	nd	nd	nd	0.03
C10:0	0.01	nd	0.02	0.03	0.04	0.33	nd	0.21	0.05
C12:0	0.08	nd	0.07	nd	0.1	1.73	1.53	1.07	nd
C14:0	0.15	0.61	0.15	0.16	0.82	8.46	1.34	1.6	0.6
C16:0	16.2	31.85	27.71	19.73	52.07	22.63	21.28	42.65	34.1
C17:0	0.22	0.43	0.2	0.28	0.77	nd	0.61	0.36	0.8
C18:0	0.92	4.11	0.8	1.34	3.14	3.81	4.38	3.04	0.88
C19:0	0.04	0.39	0.02	0.04	0.28	2.64	2.13	1.91	0.38
C20:0	0.09	7.78	nd	0.06	0.67	nd	nd	nd	nd
C12:1	0.05	nd	nd	nd	0.12	nd	nd	0.86	0.09
C14:1 ω5c	nd	nd	nd	nd	nd	nd	nd	nd	0.05
C15:1 ω5c	nd	nd	nd	nd	nd	nd	nd	nd	0.07
C15:1 ω6c	0.05	nd	0.04	0.05	nd	nd	nd	nd	nd
C15:1 ω8c	0.08	nd	nd	nd	nd	nd	nd	nd	nd
C16:1 ω7c	0.04	0.4	0.04	nd	0.09	36.66	0.83	nd	6.56
C17:1 ω6c	nd	nd	nd	nd	nd	nd	2.99	nd	nd
C17:1 ω8c	1.04	1.23	0.52	1.69	0.71	nd	1.08	1.13	1.26
C18:1 ω5c	9.99	nd	11.63	nd	nd	nd	nd	nd	nd
C18:1 ω9c	48.86	nd	43.04	58.8	nd	5.69	nd	nd	nd
C19:1 ω11c	nd	0.3	nd	0.05	nd	nd	nd	nd	nd
C20:1 ω9c	nd	nd	nd	nd	1.42	0	nd	nd	nd
C18:2 ω6c	15.31	33.79	14.58	9.99	18.98	3.47	21.17	11.5	26.94
C18:3 ω6c	nd	nd	nd	nd	2.51	nd	nd	1.36	1.79
C20:4 ω6c	nd	nd	nd	nd	nd	1.45	nd	nd	0.09
C16-C18	81.29	69.75	86.13	89.86	76.7	35.6	46.83	58.55	63.71

Percentages of FAME were calculated based on peak area of individual peaks in the GC spectrum. nd: Not detected.

Previous investigations [37,44] showed that the most common feedstocks suitable for biodiesel production were enriched in the five most common C16-C18 fatty acids, including C16:0 (palmitic acid), C18:0 (stearic acid), C18:1 (oleic acid), C18:2 (linoleic acid), and C18:3 (linolenic acid) [45]. The data in Table 4 show that the nine algal species contained considerable amounts of C16 and C18 species, ranging from 45% to 90%, except the strain*Dictyosphaerium ehrenbergianum* IL2, which contained a substantially low value at 35.6%.

Principal Component (PC) analysis was performed using FAME composition data to understand the fatty acid distribution among the isolated strains. The FA distribution is represented as PCA biplots in Figure 4. Only four fatty acids contributed to the major variations in the fatty acid composition, based on these variations, two distinctly different groups were identified, viz., an oleic acid group and a palmitic-linoleic acid group, whereas other fatty acids showed little variations. The oleic acid group included *C. vulgaris* LF5, LC8 and LC 9, which had a FA composition more similar to that of rape seed and palm oils, with C18:1 as the most abundant fatty acid [22]. Similarly the strains *C. vulgaris* LC2, *C. sorokiniana* RP1, *Micractinum* sp. IL3 and *Micractinum* sp. LC11 presented quite a similar fatty acid profile to that of soy, characterized by a high content of C18:2 and included in the palmitic-linoleic acid group.

Microalgal fatty acid profiles vary according to individual species and environmental conditions (and also during laboratory cultivation). An important aspect of strain selection and/or improvement, although until now less considered, is the FA composition of the microalgae oleaginous biomass [25,48]. The composition and structure of fatty acid methyl esters could significantly influence the fuel properties, such as degree of unsaturation and carbon chain length, and determine the fuel properties (e.g., cetane number, viscosity, cold flow, oxidative stability, and iodine value) of biodiesel [45]. However, it is difficult to say clearly state the suitability of fatty acid profiles, due to the diversity and conflicting impacts of fatty acid profiles on biodiesel properties. Therefore a highly comprehensive analysis is urgently needed to evaluate the most important biofuel properties [22].

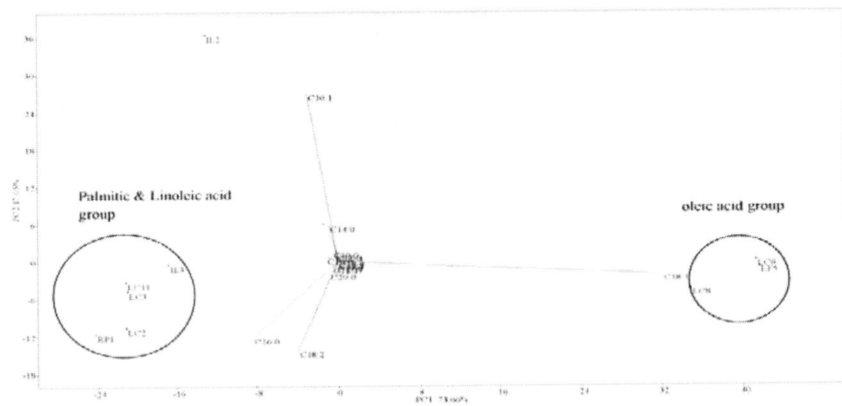

Figure 4. Principal component analysis (PCA) bi-plots for the distribution of isolated microalgal strains in relation to their FAME composition.

2.8. Predicted Fuel Properties

As shown in Table 5, eight important biodiesel properties besides average degree of unsaturation of the nine candidates were predicted. The most important fuel properties considered to assess the potential of biodiesel as substitute of diesel fuel are viscosity, cetane number (CN), density, cold filter plugging point (CFPP), oxidative stability, lubricity, ignition quality, combustion heat and cold flow [6,49].

The quality standards for diesel fuel require a minimum cetane number of 40; in the present study, the value of CN calculated for selected strains ranged from 56 to 61, which is in accordance with the standards reported as a minimum cetane number of 47 [45] why different values? High cetane value is one of the important fuel property indicators of better combustion, low nitrous oxide (N_2O) emissions, less occurrence of knocking and easier engine start-up [46,50].

Iodine value in the present study was found to be less than 80.39 g I_2/100 g in all strains, which satisfies the European biodiesel standards (Table 5). Higher iodine values may result in the polymerization of glycerides and deposition of lubricant in the engine [51]. The low degree of unsaturation found in these strains is crucial for the overall performance of diesel engines and an encouraging feature for biofuel production.

The melting point of saturated fatty acids is always higher than that of unsaturated fatty acid compounds. When the oil contains mostly saturated FA ester molecules, crystallization may occur at temperatures within the normal engine operation range [51], which results in poor CFPP properties. Biodiesel rich in palmitic and stearic acid methyl esters has a tendency to present a poor CFPP (equivalent to a higher plugging point temperature), because when a liquid biodiesel is cooled, these FAME precipitate first [46]. In the present study, the levels of palmitic and stearic acid (Table 4) were low (below 35%), except for *Chlorella sorokiniana* RP1, *C. vulgaris* LC2 and*Micractinum* sp. LC11 (55.2%, 36.0% and 45.7%, respectively). These low stearic and palmitic acid values may have contributed to the lower CFPP temperatures of the majority of the studied strains. The CFPP values estimated for biodiesel from the strains studied in the present work ranged from 6.91 °C (*Chlorella sorokiniana*RP1) to −9.66 °C (*Chlorella vulgaris* LF5). According to a previous report, the CFPP values obtained from microalgal oils vary from −12.3 to 20.8 °C [51] and LCSF for the microalgal isolates were in accordance with the international standards revealing the flow performance of biodiesel at low temperatures [25] without greatly affecting the cold flow properties of biodiesel [50].

3. MATERIALS AND METHODS

3.1. Sampling

Microalgae were sampled from four different locations: Lacul Ciucas (LC) in Romania, Inner Lake (IL), Tihany, Lake Velencei surrounded by a reed zone (RP) and from Lake Feneketlen (LF) in Hungary (see sampling site details in Table 1). The predominant criterion for sample collection was the presence of abundant microalgae in the water [25]. Temperature, pH and conductivity of each sample were determined *in situ* with a Multi-Line P4 meter (Wissenschaftlich-Technische Werkstätten, Weilheim, Germany) and the samples were transported to the laboratory in a thermo box in dark conditions.

Table 5. Comparison of nine properties of biodiesel from microalgal oil, biodiesel, and American Society for Testing and Materials (ASTM) biodiesel, European (EN) biodiesel standard.

Property	LF5	LC2	LC8	LC9	RP1	IL2	IL3	LC3	LC11	Biodiesel	US (ASTM D6751-08)	Europe (EN 14214)
Kinematic viscosity 40 °C ($mm^2 s^{-1}$)	4.63	4.77	4.67	4.69	4.90	4.91	5.04	5.07	4.95	4–5	1.9–6.0	3.5–5.0
Specific gravity (kg^{-1})	0.878	0.876	0.877	0.877	0.875	0.875	0.874	0.873	0.875	0.87–0.89	0.85–0.9	-
Cloud point (°C)	7.84	10.77	8.78	9.18	13.58	13.85	16.66	17.19	14.65	-	-	-
Cetane number	56.81	58.27	57.27	57.47	59.68	59.81	61.21	61.48	60.21	45–55	Minimum 47	Minimum 51–Maximum 120
Iodine value (g Iodine/100 g)	80.39	64.03	75.18	72.95	48.41	46.92	31.30	28.33	42.46	-	-	Maximum 120
HHV (MJ/kg)	40.14	39.75	40.01	39.96	39.38	39.34	38.97	38.90	39.24	-	-	-
Average Unsaturation	0.91	0.69	0.84	0.81	0.48	0.46	0.25	0.20	0.40	-	-	-
Long-chain saturated factor (LCFF)	2.17	13.02	3.171	2.703	7.447	4.188	4.318	5.785	3.85	-	-	-
Cold Filter Plugging Point (CFPP)	−9.66	2.43	−6.51	−7.98	6.91	−3.31	−2.91	1.69	−4.38	-	-	−5 to −13

3.2. Isolation and Cultivation of Microalgal Strains

Sample processing started within 3–6 h after sampling. The samples were filtered through double-layered blotting sheets to remove any solid contaminations and debris and 100 µL of filtrate were then transferred onto the BG11 (Blue Green Medium) solid culture media [52] and incubated in a light chamber at 20 ± 2 °C for two weeks. After growth, different colonies were picked, cultivated photoautotrophically in 125 mL Erlenmeyer flasks containing 60 mL of BG11 medium [53] and incubated in a light mounted shaker at 20 ± 2 °C, with shaking at 120 RPM and a light intensity of 21.2 W/m^2 using a photoperiod of 12 h light:

dark, in CO_2 limited condition for 21 days. Axenicity of the isolates was confirmed by cultivation on LB plates and observation with a light microscope following the method [25].

3.3. Morphological Investigations

During the isolation process, algal strains were preliminarily identified by observation of morphological characteristics under a light microscope and using botanical approaches as described [54].

3.4. Molecular Characterization

Taxonomic determination was further confirmed by sequencing the 18S ribosomal RNA gene. Genomic DNA was extracted from the isolates with a DNeasy Plant Mini Kit (Qiagen, Hilden, Germany). PCR was performed with primer pairs Euk328f-Chlo02R [55,56] as described by Somogyi *et al.* [57]. Reactions were carried out in a final volume of 50 μL using 2 μL genomic DNA, 0.2 mM of each deoxynucleotide, 1 U LC *Taq* (*Thermus aquaticus*) DNA polymerase (Fermentas, Vilnius, Lithuania), 1 × PCR buffer (Fermentas), 2 mM $MgCl_2$, 0.325 mM of each primer and 20 μg of BSA (Bovine Serum Albumin) (Fermentas). The thermal profile consisted of a first denaturing step at 98 °C for 3 min, followed by 32 amplification cycles of denaturation at 94 °C for 45 s, annealing at 55 °C for 60 s, elongation at 72 °C for 1.30 min and a final extension step at 72 °C for 10 min. PCR product purification, sequencing reaction and capillary electrophoresis were performed by the LGC Genomics GmbH (Berlin, Germany). Chromatograms were corrected manually with Chromas 1.45 software (Technelysium Pty Ltd., South Brisbane, Australia). The generated sequences were compared to the GenBank nucleotide database using the Blast program [58]. The obtained 18S rRNA gene sequences were submitted to GenBank under the accession numbers given in Table 2. Phylogenetic analysis was performed using the Molecular Evolutionary Genetics Analysis v5 (MEGA5) software [59] using an alignment created with SINA Aligner [60].

3.5. Nile Red Staining of Neutral Lipid Droplets

To visualize TAG accumulation in microalgae, strains were stained with lipid-sensitive Nile Red fluorescent dye, observed under fluorescence microscope for detecting the presence of intracellular lipid droplets [61]. Staining was carried out on fixed (1.5% glycerol for 5 min) cells [33]. The dye was then added directly to the preparation to effect 1:50 / 1:100 dilutions and incubated for 5–10 min in the dark at room temperature [33]. The fluorescence intensity of the stained microalgal cells was measured at an excitation and emission wavelength of 450 and 510 nm, respectively, using a Nikon 80i (Nikon, Tokyo, Japan) epifluorescence microscope equipped with a camera Q-Imaging Micropublisher 3.3 RTV (Real Time Viewing), and the software used to capture images was the Image-Pro Plus v6.0 (Media Cybernetics, Inc., Bethesda, MD, USA).

3.6. Determination of Total Lipids

The Bligh and Dyer method [62] for lipid extraction was used in this study with slight modifications. The harvested cells were disrupted using ultra sonic bath (Equitron, Mumbai, India) at frequency of 153 KHz for 1 min. All samples were extracted with 3 mL of chloroform/methanol at a ratio of 1:2 volumes for volume (V/V) by vortexing (1 min) and centrifugation at 8000 rpm for 15 min at room temperature (RT). The supernatants were collected and residues were re-extracted thrice with 2 mL of chloroform/methanol (1/1, V/V) by centrifugation as stated above. All the supernatants were pooled together, filtered with Whatman No. 1 filter paper (Whatman Inc., Clifton, NJ, USA), and washed with 2 mL of milli-Q water, followed by centrifugation using glass centrifuge tubes (model T32c Janetzki, Olympus, Japan) at 8000 rpm for 5 min. The lower organic phases were collected and evaporated to dryness. Samples were resuspended in chloroform for further analysis.

3.7. Fatty Acid Methyl Esters (FAME) Determination

Fatty acid methyl esters (FAME) present in the algal lipids were determined following the technique proposed by Freedman *et al.* [63] with some modifications. The conversion of lipids into FAME was performed by acid-catalyzed methylation. The dried lipid (50 mg) extract was added to 15–20 mL of 2% H_2SO_4 in methanol and refluxed for about 4 h 30 min. After complete conversion as monitored by thin layer chromatography (TLC), the solvent was partially removed and the remaining mixture was extracted with ethyl acetate (20 mL), and the combined ethyl acetate layers were washed with water until pH was neutral. The ethyl acetate extract was dried over anhydrous Na_2SO_4 and evaporated under reduced pressure on a rotary evaporator (Büchi R-200, Flawil, Switzerland) to recover the FAME fraction. Methyl esters of fatty acids were analyzed using a gas chromatograph (GC-6890N, Agilent, Santa Clara, CA, USA) equipped with a flame ionization detector DB225 capillary column (30 m × 0.25 mm I.D.; 0.25 μm—Agilent Technologies). The initial oven temperature was maintained at 160 °C for 2 min with a sequential increase to 180 °C at 6 °C min^{-1} for 2 min and 230 °C at 4 °C min^{-1}. The final oven temperature was maintained at 230 °C for 10 min. Nitrogen was used as the carrier gas with a flow rate of 1.5 mL min^{-1}. The injector and FID temperatures were set at 230 and 250 °C respectively, while a split ratio of 50:1 was maintained for the analysis. Heptadecanoic acid was used as the internal standard for quantitative analysis. The components were identified by comparing their retention times and fragmentation patterns with those of the standards and expressed as a percentage of the total fatty acids identified in the oil following the description [64].

3.8. Calorimetric Determination of Total Protein

Total proteins were analysed by the Folin phenol reagent method as described by Lowry *et al.* [65]. Samples were prepared by boiling resuspended cells at 100 °C for 10 min in the presence of 1 N (Normality) NaOH. Aliquots (1 mL) were subjected to protein determination. The calibration curve was obtained using bovine serum albumin (BSA) as a reference standard.

3.9. Determination of Total Carbohydrates

Total soluble carbohydrates were analysed by the phenol-sulphuric acid method as described by Dubois *et al.*[66]. The calibration curve was obtained using D-glucose (Hi-Media Laboratories, Mumbai, India) as a reference standard.

3.10. Bioprospecting of Biodiesel Properties Based on FAME Profiles

Predictive models based on FA composition were used in this study for the calculation of critical biodiesel properties, otherwise it would require large amounts of diesel and specialized instrumentation, which are not always readily available. In recent studies, many equations based on FA composition have been built to predict the properties of biodiesel [22,51,67].

In this work, the equations of Hoekman *et al.* [22] were selected to predict the properties of biodiesel, since the calculated values using the equation are more closer to the measured values from biodiesel [37,44].

The Average Degree of Unsaturation (ADU) calculated from compositional profiles of fatty acid were calculated using Equation (1):

$$ADU = \sum M \times Yi \quad (1)$$

where M is the number of carbon–carbon double bonds in FA constituent and Yi is the mass fraction of each FA constituent respectively.

The relationships between average degree of unsaturation and biodiesel properties including kinematic viscosity, specific gravity, cloud point, cetane number, iodine value and higher heating value (HHV) are as shown in Equations (2)–(7) [22]:

$$\text{Kinematic viscosity} = -0.6313X + 5.2065 \quad (2)$$

$$\text{Specific gravity} = 0.0055X + 0.8726 \quad (3)$$

$$\text{Cloud point} = -13.356X + 19.994 \quad (4)$$

$$\text{Cetane number} = -6.6684X + 62.876 \quad (5)$$

$$\text{Iodine value} = 74.373X + 12.71 \quad (6)$$

$$HHV = 1.7601X + 38.534 \quad (7)$$

where X is the ADU.

The long-chain saturated factor (LCSF) directly used to calculate Cold Filter Plugging Point (CFPP) was also estimated through Equations (8) and (9). These two factors are both related to chain saturation and length of FAME [21]:

$$LCSF = (0.1 \times C_{16}) + (0.5 \times C_{18}) + (1 \times C_{20}) + (1.5 \times C_{22}) + (2 \times C_{24}) \quad (8)$$

$$CFPP = (3.1417 \times LCSF) - 16.477 \quad (9)$$

3.11. Statistical Analysis

The experiments adopted for the estimation of total protein, carbohydrate and lipids were carried out in triplicate and data are expressed as mean ± standard deviation. Principal component analysis

(PCA) for the fatty acids was accomplished to explore the underlying inter-relationships between them. PCA bi-plots were generated using PAST (paleontological statistics) 2.06 statistical software [68].

4. CONCLUSIONS

Currently, screening and evaluation of potential green algal strains *viz* biomass productivity, lipid cell content, fatty acids and fuel properties are some of the key parameters that determine the economic feasibility of algal oils for biodiesel production. In this work, nine *Chlorophyceae* strains isolated from freshwater and soda lakes from Central Europe were compared according to their biomass and lipid productivities. The total lipids ratios in dry biomass varied from as low as 8.9% to ratios as high as 42.1%. The highest lipid yields were observed from the *C. vulgaris* LC8 strain with a favorable C16-C18 fatty acid profile of 77.38%, as well as desirable biodiesel properties like a high cetane number (57.3), low viscosity (4.67 mm^2/s), lower iodine number (75.2 g I_2/100 g), relative cloud point (8.8 °C) and negative cold filter plugging point (−6.5 °C). *C. vulgaris* LC8 presented one of the best combinations of desirable traits, which makes it a good candidate for further assessment for biofuel production. Results suggest that the adequate fatty acid composition of microalgal oil and the lipid productivity of strains must be priority criteria for strain selection, to make the algae-based biodiesel industry viable.

ACKNOWLEDGMENTS

This work was supported by the Hungarian Scientific Research Fund, grant number PD 105407. Furthermore, R.S. was supported by the Balassi Institute-Hungarian Scholarship Board and T.F. was supported by the János Bolyai Research Scholarship of the Hungarian Academy of Sciences.

AUTHOR CONTRIBUTIONS

R.S. (corresponding author) conducted the experiments and analyzed the data and wrote the manuscript. E.S. supervised the study in India while T.F. supervised the study in Hungary. T.T. helped in Fatty Acid analysis. T.S. and M.T. coordinated in the writing of the manuscript. All authors read and approved the final manuscript.

REFERENCES

1. Chisti, Y. Biodiesel from microalgae. *Biotechnol. Adv.* **2007**, *25*, 294–306.
2. Huang, G.; Chen, F.; Wei, D.; Zhang, X.; Chen, G. Biodiesel production by microalgal biotechnology. *Appl. Energy* **2010**, *87*, 38–46.
3. Chen, F. Microalgae and their biotechnological potential. *J. Biotechnol.* **2008**, *136*, S521.]
4. Buijks, J. Future of Algae Based Biodiesel Production in The Netherlands. Master's Thesis, University of Utrecht, Utrecht, The Netherlands, July 2013.
5. Scott, S.A.; Davey, M.P.; Dennis, J.S.; Horst, I.; Howe, C.J.; Lea-Smith, D.J.; Smith, A.G. Biodiesel from algae: Challenges and prospects. *Curr. Opin. Biotechnol.* **2010**, *21*, 277–286.
6. Knothe, G. Biodiesel and renewable diesel: A comparison. *Prog. Energy Combust. Sci.* **2010**, *36*, 364–373.
7. Lundquist, T.; Woertz, I.; Quinn, N.; Benemann, J. *A Realistic Technology and Engineering Assessment of Algae Biofuel Production*; University of California: Berkeley, CA, USA, 2010.
8. Woertz, I.; Feffer, A.; Lundquist, T.; Nelson, Y. Algae grown on dairy and municipal wastewater for simultaneous nutrient removal and lipid production for biofuel feedstock. *J. Environ. Eng.* **2009**, *135*, 1115–1122.
9. Rawat, I.; Kumar, R.R.; Mutanda, T.; Bux, F. Dual role of microalgae: Phycoremediation of domestic wastewater and biomass production for sustainable biofuels production. *Appl. Energy* **2011**, *88*, 3411–3424.
10. Yun, Y.-S.; Lee, S.B.; Park, J.M.; Lee, C.-I.; Yang, J.-W. Carbon Dioxide Fixation by Algal Cultivation Using Wastewater Nutrients. *J. Chem. Technol. Biotechnol.* **1997**, *69*, 451–455.
11. Pittman, J.K.; Dean, A.P.; Osundeko, O. Bioresource technology the potential of sustainable algal biofuel production using wastewater resources. *Bioresour. Technol.* **2011**, *102*, 17–25.
12. Park, D.H.; Ruy, H.W.; Lee, K.Y.; Kang, C.H.; Kim, T.H.; Lee, H.Y. The production of hydrocarbons from photoautotrophic growth of dunaliella salina 1650. *Appl. Biochem. Biotechnol.* **1998**, *70*, 739–746.

13. Spolaore, P.; Joannis-cassan, C.; Duran, E.; Isambert, A. Commercial Applications of Microalgae. *J. Biosci. Bioeng.* **2006**, *101*, 87–96.
14. Mata, T.M.; Martins, A.A.; Caetano, N.S. Microalgae for biodiesel production and other applications: A review. *Renew. Sustain. Energy Rev.* **2010**, *14*, 217–232.
15. Abomohra, A.E.F.; Wagner, M.; El-Sheekh, M.; Hanelt, D. Lipid and total fatty acid productivity in photoautotrophic fresh water microalgae: Screening studies towards biodiesel production. *J. Appl. Phycol.***2013**, *25*, 931–936.
16. Selvarajan, R.; Velu, P.; Sanniyasi, E. Molecular Characterization and Fatty Acid Profiling Of Indigenous Microalgae Species with Potential For Biofuel Production In Tamil Nadu India. *Golden Res. Thoughts* **2014**,*3*, 1–9.
17. Abou-shanab, R.A.I.; Hwang, J.; Cho, Y.; Min, B.; Jeon, B. Characterization of microalgal species isolated from fresh water bodies as a potential source for biodiesel production. *Appl. Energy* **2011**, *88*, 3300–3306.
18. Sivakumar, G.; Jeong, K.; Lay, J.O. Bioprocessing of stichococcus bacillaris strain siva 2011. *Biotechnol. Biofuels* **2014**, *7*, 62.
19. Sivakumar, G.; Xu, J.; Thompson, R.W.; Yang, Y.; Randol-Smith, P.; Weathers, P.J. Integrated green algal technology for bioremediation and biofuel. *Bioresour. Technol.* **2012**, *107*, 1–9.
20. Yun, H.S.; Lee, H.; Park, Y.T.; Ji, M.K.; Kabra, A.N.; Jeon, C.; Jeon, B.H.; Choi, J. Isolation of novel microalgae from acid mine drainage and its potential application for biodiesel production. *Appl. Biochem. Biotechnol.* **2014**, *173*, 2054–2064.
21. Talebi, A.F.; Mohtashami, S.K.; Tabatabaei, M.; Tohidfar, M.; Bagheri, A.; Zeinalabedini, M.; Hadavand Mirzaei, H.; Mirzajanzadeh, M.; Malekzadeh Shafaroudi, S.; Bakhtiari, S. Fatty acids profiling: A selective criterion for screening microalgae strains for biodiesel production. *Algal Res.* **2013**, *2*, 258–267.
22. Hoekman, S.K.; Broch, A.; Robbins, C.; Ceniceros, E.; Natarajan, M. Review of biodiesel composition, properties, and specifications. *Renew. Sustain. Energy Rev.* **2012**, *16*, 143–169.
23. Sydney, E.B.; da Silva, T.E.; Tokarski, A.; Novak, A.C.; de Carvalho, J.C.; Woiciecohwski, A.L.; Larroche, C.; Soccol, C.R. Screening of microalgae with potential for biodiesel production and nutrient

removal from treated domestic sewage. *Appl. Energy* **2011**, *88*, 3291–3294.

24. Rawat, I.; Ranjith Kumar, R.; Mutanda, T.; Bux, F. Biodiesel from microalgae: A critical evaluation from laboratory to large scale production. *Appl. Energy* **2013**, *103*, 444–467.
25. Do Nascimento, M.; Ortiz-Marquez, J.C.F.; Sanchez-Rizza, L.; Echarte, M.M.; Curatti, L. Bioprospecting for fast growing and biomass characterization of oleaginous microalgae from South-Eastern Buenos Aires, Argentina. *Bioresour. Technol.* **2012**, *125*, 283–290.
26. Abou-shanab, R.A.I.; Matter, I.A.; Kim, S.; Oh, Y.; Choi, J.; Jeon, B. Characterization and identification of lipid-producing microalgae species isolated from a freshwater lake. *Biomass Bioenergy* **2011**, *35*, 3079–3085.
27. Hu, Q.; Sommerfeld, M.; Jarvis, E.; Ghirardi, M.; Posewitz, M.; Seibert, M.; Darzins, A. Microalgal triacylglycerols as feedstocks for biofuel production: Perspectives and advances. *Plant J.* **2008**, *54*, 621–639.
28. Pereira, H.; Barreira, L.; Custódio, L.; Alrokayan, S.; Mouffouk, F.; Varela, J.; Abu-Salah, K.M.; Ben-Hamadou, R. Isolation and fatty acid profile of selected microalgae strains from the red sea for biofuel production. *Energies* **2013**, *6*, 2773–2783.
29. Pereira, H.; Barreira, L.; Mozes, A.; Florindo, C.; Polo, C.; Duarte, C.V.; Custódio, L.; Varela, J. Microplate-based high throughput screening procedure for the isolation of lipid-rich marine microalgae. *Biotechnol. Biofuels* **2011**, *4*, 1–12.
30. Matsunaga, T.; Matsumoto, M.; Maeda, Y.; Sugiyama, H.; Sato, R.; Tanaka, T. Characterization of marine microalga, *Scenedesmus* sp. strain JPCC GA0024 toward biofuel production. *Biotechnol. Lett.* **2009**, *31*, 1367–1372.
31. Moazami, N.; Ranjbar, R.; Ashori, A.; Tangestani, M. Biomass and lipid productivities of marine microalgae isolated from the persian gulf and the qeshm island. *Biomass Bioenergy* **2011**, *35*, 1935–1939.
32. Tripathi, R.; Singh, J.; Thakur, I.S. Characterization of microalga *Scenedesmus* sp. ISTGA1 for potential CO_2sequestration and biodiesel production. *Renew. Energy* **2015**, *74*, 774–781.
33. Doan, T.T.Y.; Obbard, J.P. Improved nile red staining of *Nannochloropsis* sp. *J. Appl. Phycol.* **2011**, *23*, 895–901.

34. Brennan, L.; Owende, P. Biofuels from microalgae—A review of technologies for production, processing, and extractions of biofuels and co-products. *Renew. Sustain. Energy Rev.* **2010**, *14*, 557–577.
35. Liu, J.; Huang, J.; Chen, F. Microalgae as Feedstocks for Biodiesel Production. In *Biodiesel-Feedstocks and Processing Technologies*; Stoytcheva, M.C., Ed.; INTECH: Rijeka, Croatia, 2011; pp. 58–78.
36. Phukan, M.M.; Chutia, R.S.; Konwar, B.K.; Kataki, R. Microalgae Chlorella as a potential bio-energy feedstock. *Appl. Energy* **2011**, *88*, 3307–3312.
37. Song, M.; Pei, H.; Hu, W.; Ma, G. Evaluation of the potential of 10 microalgal strains for biodiesel production. *Bioresour. Technol.* **2013**, *141*, 245–251.
38. Illman, A.M.; Scragg, A.H.; Shales, S.W. Increase in Chlorella strains calorific values when grown in low nitrogen medium. *Enzyme Microb. Technol.* **2000**, *27*, 631–635.
39. Feng, D.; Chen, Z.; Xue, S.; Zhang, W. Increased lipid production of the marine oleaginous microalgae*Isochrysis* zhangjiangensis (Chrysophyta) by nitrogen supplement. *Bioresour. Technol.* **2011**, *102*, 6710–6716.
40. Chen, C.Y.; Zhao, X.Q.; Yen, H.W.; Ho, S.H.; Cheng, C.L.; Lee, D.J.; Bai, F.W.; Chang, J.S. Microalgae-based carbohydrates for biofuel production. *Biochem. Eng. J.* **2013**, *78*, 1–10.
41. Trzcinski, A.P.; Hernandez, E.; Webb, C. A novel process for enhancing oil production in algae biorefineries through bioconversion of solid by-products. *Bioresour. Technol.* **2012**, *116*, 295–301.
42. Wang, Y.; Guo, W.; Lo, Y.; Chang, J.; Ren, N. Characterization and kinetics of bio-butanol production with Clostridium acetobutylicum ATCC824 using mixed sugar medium simulating microalgae-based carbohydrates. *Biochem. Eng. J.* **2014**, *91*, 220–230.
43. Pulz, O.; Gross, W. Valuable products from biotechnology of microalgae. *Appl. Microbiol. Biotechnol.* **2004**,*65*, 635–648.
44. Ma, Y.; Wang, Z.; Yu, C.; Yin, Y.; Zhou, G. Evaluation of the potential of 9 Nannochloropsis strains for biodiesel production. *Bioresour. Technol.* **2014**, *167*, 503–509.
45. Knothe, G. Analyzing biodiesel: Standards and other methods. *J. Am. Oil Chem. Soc.* **2006**, *83*, 823–833.

46. Knothe, G. "Designer" biodiesel: Optimizing fatty ester composition to improve fuel properties. *Energy Fuels* **2008**, *22*, 1358–1364.
47. Stournas, S.; Lois, E.; Serdari, A. Effects of fatty acid derivatives on the ignition quality and cold flow of diesel fuel. *J. Am. Oil Chem. Soc.* **1995**, *72*, 433–437.
48. Mutanda, T.; Ramesh, D.; Karthikeyan, S.; Kumari, S.; Anandraj, A.; Bux, F. Bioprospecting for hyper-lipid producing microalgal strains for sustainable biofuel production. *Bioresour. Technol.* **2011**, *102*, 57–70.
49. Georgianna, D.R.; Mayfield, S.P. Exploiting diversity and synthetic biology for the production of algal biofuels. *Nature* **2012**, *488*, 329–335.
50. Mandotra, S.K.; Kumar, P.; Suseela, M.R.; Ramteke, P.W. Fresh water green microalga Scenedesmus abundans: A potential feedstock for high quality biodiesel production. *Bioresour. Technol.* **2014**, *156*, 42–47.
51. Francisco, É.C.; Neves, D.B.; Jacob-Lopes, E.; Franco, T.T. Microalgae as feedstock for biodiesel production: Carbon dioxide sequestration, lipid production and biofuel quality. *J. Chem. Technol. Biotechnol.* **2010**, *85*, 395–403.
52. Stanier, R.Y.; Kunisawa, R.; Mandel, M.; Cohen-Bazire, G. Purification and properties of unicellular blue-green algae (order Chroococcales). *Bacteriol. Rev.* **1971**, *35*, 171–205.
53. Rippka, R.; Deruelles, J.; Waterbury, J.B.; Herdman, M.; Stanier, R.Y. Generic assignments, strain histories and properties of pure cultures of cyanobacteria. *J. Gen. Microbiol.* **1979**, *111*, 1–61.
54. Whitton, P.B.A.; John, D.M.; Kelly, M.G.; Haworth, E.Y.; Tindall, C.I. *The Freshwater Algal Flora of the British Isles an Identification Guide to Freshwater and Terrestrial Algae*; Cambridge University Press: Cambridge, UK, 2003.
55. Zhu, F.; Massana, R.; Not, F.; Marie, D.; Vaulot, D. Mapping of picoeucaryotes in marine ecosystems with quantitative PCR of the 18S rRNA gene. *FEMS Microbiol. Ecol.* **2005**, *52*, 79–92.
56. Moon-van der Staay, S.Y.; van der Staay, G.W.M.; Guillou, L.; Vaulot, D.; Claustre, H.; Medlin, L.K. Abundance and diversity of prymnesiophytes in the picoplankton community from the equatorial Pacific Ocean inferred from 18S rDNA sequences. *Limnol. Oceanogr.* **2000**, *45*, 98–109.

57. Somogyi, B.; Felföldi, T.; Solymosi, K.; Flieger, K.; Márialigeti, K.; Böddi, B.; Vörös, L. One step closer to eliminating the nomenclatural problems of minute coccoid green algae: *Pseudochloris wilhelmii, gen. et sp. nov.* (Trebouxiophyceae, Chlorophyta). *Eur. J. Phycol.* **2013**, *48*, 427–436.
58. Altschul, S.F.; Madden, T.L.; Schäffer, A.A.; Zhang, J.; Zhang, Z.; Miller, W.; Lipman, D.J. Gapped BLAST and PSI-BLAST: A new generation of protein database search programs. *Nucleic Acids Res.* **1997**, *25*, 3389–3402.
59. Tamura, K.; Peterson, D.; Peterson, N.; Stecher, G.; Nei, M.; Kumar, S. MEGA5: Molecular evolutionary genetics analysis using maximum likelihood, evolutionary distance, and maximum parsimony methods.*Mol. Biol. Evol.* **2011**, *28*, 2731–2739.
60. Pruesse, E.; Peplies, J.; Glöckner, F.O. SINA: Accurate high-throughput multiple sequence alignment of ribosomal RNA genes. *Bioinformatics* **2012**, *28*, 1823–1829.
61. Elsey, D.; Jameson, D.; Raleigh, B.; Cooney, M.J. Fluorescent measurement of microalgal neutral lipids. *J. Microbiol. Methods* **2007**, *68*, 639–642.
62. Bligh, E.G.; Dyer, W.J. A rapid method of total lipid extraction and purification. *Can. J. Biochem. Physiol.* **1959**, *37*, 911–917.
63. Freedman, B.; Butterfield, R.O.; Pryde, E.H. Transesterification kinetics of soybean oil. *J. Am. Oil Chem. Soc.* **1986**, *63*, 1375–1380.
64. Sahu, A.; Pancha, I.; Jain, D.; Paliwal, C.; Ghosh, T.; Patidar, S.; Bhattacharya, S.; Mishra, S. Fatty acids as biomarkers of microalgae. *Phytochemistry* **2013**, *89*, 53–58.
65. Lowry, O.H.; Rosebrough, N.J.; Farr, A.L.; Randall, R.J. Protein measurement with the folin phenol reagent.*J. Biol. Chem.* **1951**, *193*, 265–275.
66. DuBois, M.; Gilles, K.A.; Hamilton, J.K.; Rebers, P.A.; Smith, F. Colorimetric method for determination of sugars and related substances. *Anal. Chem.* **1956**, *28*, 350–356.
67. Nascimento, I.A.; Marques, S.S.I.; Cabanelas, I.T.D.; Pereira, S.A.; Druzian, J.I.; de Souza, C.O.; Vich, D.V.; de Carvalho, G.C.; Nascimento, M.A. Screening microalgae strains for biodiesel production: Lipid productivity and estimation of fuel quality based on fatty acids profiles as aelective criteria. *Bioenergy Res.* **2013**, *6*, 1–13.

68. Hammer, Ø.; Harper, D.A.T.; Ryan, P.D. Paleontological statistics software package for education and data analysis. *Palaeontol. Electron.* **2001**, *4*, 9–18.

8
CHAPTER

BIOMASS FEEDSTOCK PRE-PROCESSING – PART 2: DENSIFICATION

Lope Tabil[1], Phani Adapa[1] and Mahdi Kashaninejad[2]

[1]*Department of Chemical and Biological Engineering, University of Saskatchewan, Canada*

[2]*Department of Food Science & Technology, Gorgan University of Agricultural Sciences and Natural Resources Gorgan, Iran*

1. INTRODUCTION

1.1. The Need for Densification

Agricultural biomass residues have the potential for the sustainable production of bio-fuels and to offset greenhouse gas emissions (Campbell et al., 2002; Sokhansanj et al., 2006). Straw from crop production and agricultural residues existing in the waste streams from commercial crop processing plants have little inherent value and have traditionally constituted a disposal problem. In fact, these residues represent an abundant, inexpensive and readily available source of renewable lignocellulosic biomass (Liu et al., 2005). New methodologies need to be developed to process the biomass making it suitable feedstock for bio-fuel production. In addition, some of the barriers in the economic use of agricultural crop residue are the variable quality of the residue, the cost of collection, and problems in transportation and storage (Bowyer and Stockmann, 2001; Sokhansanj et al., 2006).

In order to reduce industry's operational cost as well as to meet the requirement of raw material for biofuel production, biomass must be processed and handled in an efficient manner. Due to its high moisture content, irregular shape and size, and low bulk density, biomass is very difficult to handle, transport, store, and utilize in its original form (Sokhansanj et al., 2005). Densification of biomass into durable compacts is an effective solution to these problems and it can reduce material waste. Densification can increase the bulk density of biomass from an initial bulk density of 40-200 kg/m^3 to a final compact density of 600-1200 kg/m^3 (Adapa et al., 2007; Holley, 1983; Mani et al., 2003;McMullen et al., 2005; Obernberger and Thek, 2004). Biomass can be compressed and stabilized to 7–10 times densities of the standard bales by the application of pressures between 400–800 MPa during the densification process (Demirbas and Sahin, 1998). Because of their uniform shape and size, densified products may be easily handled using standard handling and storage equipment, and they can be easily adopted in direct-combustion or co-firing with coal, gasification, pyrolysis, and utilized in other biomass-based conversions (Kaliyan and Morey, 2006a) such as biochemical processes. Upon densification, many agricultural biomass materials, especially those from straw and stover, result in a poorly formed pellets or compacts that are more often dusty, difficult to handle and costly to manufacture. This is caused by lack of complete understanding on the natural binding characteristics of the components that make up biomass (Sokhansanj et al., 2005).

1.2. Fuel Pellet Quality Parameters

The quality of fuel pellet is usually assessed based on its density and durability. High density of pellet represents higher energy per unit volume of material, while durability is the resistance of pellets to withstand various shear and impact forces applied during handling and transportation. High bulk density increases storage and transport capacity of pellets. Since feeding of boilers and gasifiers generally is volume-dependent, variations in bulk density should be avoided (Larsson et al., 2008). A bulk density of 650 kg/m^3 is stated as design value for wood pellet producers (Obernberger and Thek, 2004). Low durability of pellets results in problems like disturbance within pellet feeding systems, dust emissions, and an increased risk of fire and explosions during pellet handling and storage (Temmerman et al., 2006). Other quality factors of biomass for thermo-chemical conversion include (FAO, 2011;Rajvanshi, 1986):

- *Energy content:* The choice of a biomass for energy conversion will in part be decided by its heating value. The method of measurement of the biomass energy content will influence the estimate of efficiency of a given gasifier. The only realistic way of presenting fuel heating values for gasification purposes is to give lower heating values (excluding the heat of condensation of the water produced) on an ash inclusive basis and with specific reference to the actual moisture content of the fuel.

- *Moisture content:* High moisture contents reduce the thermal efficiency since heat is used to drive off the water and consequently this energy is not available for the reduction reactions and for converting thermal energy into chemical bound energy in the gas. Therefore, high moisture contents result in low gas heating values during thermo-chemical processes.

- *Volatile matter:* The amount of volatiles in the feedstock determines the necessity of special measures (either in design of the gasifier or in the layout of the gas cleanup train) in order to remove tars from the product gas in engine applications.

- *Ash content and slagging characteristics:* The mineral content in the biomass that remains in oxidation form after complete combustion is usually called ash. The ash content of a fuel and the ash composition have a major impact on trouble free operation of a gasifier or a burner. Slagging or clinker formation in the reactor, caused by melting and agglomeration of ashes, at the best will greatly add to the amount of labour required to operate the gasifier. If no special measures are taken, slagging can lead to excessive tar formation and/or complete blocking of the reactor.

- *Reactivity:* The reactivity is an important factor determining the rate of reduction of carbon dioxide to carbon monoxide in a gasifier. Reactivity depends in the first instance on the type of fuel. For example, it has been observed that fuels such as wood, charcoal and peat are far more reactive than coal.

- *Size and size distribution:* Low bulk density feedstock may cause flow problems in the gasifier or burner as well as an inadmissible pressure drop over the reduction zone and a high proportion of dust in the gas. Large pressure drops will lead to reduction of the gas load, resulting in low temperatures and tar production. Excessively large sizes of particles or pieces give rise to reduction in reactivity of the fuel, resulting in start-up problems and poor gas quality, and to transport problems through the equipment. A large range in size distribution of the feedstock will generally aggravate the above phenomena. Too large particle sizes can cause gas channelling problems. Fluidized bed gasifiers are normally able to handle fuels with particle diameters varying between 0.1 and 20 mm (FAO, 2011).

- *Bulk density:* Fuels with high bulk density are advantageous because they represent a high energy-for-volume value. Consequently, these fuels need less bunker space for a given refuelling time. Low bulk density fuels sometimes give rise to insufficient flow under gravity, resulting in low gas heating values and ultimately in burning of the char in the reduction zone. Inadequate bulk densities can be improved by briquetting or pelletizing.

All of the abovementioned biomass properties could be altered by subjecting raw biomass to various processing methods and forming composites. Before choosing a gasifier, it is important to ensure that the individual biomass meets the requirements of the gasifier or that it can be treated to meet these requirements.

1.3. Effect of Moisture Content on Pellet Quality

The moisture in biomass both acts as a facilitator of natural binding agents and a lubricant (Kaliyan and Morey, 2006a). Many studies have indicated that the production of high quality pellets is possible only if the moisture content of the feed is between 8 and 12% (wb). Moisture contents above or below this range would lead to lower quality pellets (Hill and Pulkinen, 1988; Kashaninejad et al., 2011; Li and Liu, 2000; Obernberger and Thek, 2004; Shaw and Tabil, 2007). In general, an increase in moisture content from 10 to 44% could result in up to 30-40% decrease in pellet densities of biomass (Chancellor, 1962; Grover and Mishra, 1996; Gustafson and Kjelgaard, 1963; Kaliyan and Morey, 2006a; Mani et al., 2002 and 2006b; Smith et al., 1977). However, the percentage decrease in density depends on the type of biomass. Therefore, a moisture content of 10% (w.b.) is considered as optimal moisture content to obtain high density and durability pellets.

1.4. Effect of Grind Size on Pellet Quality

In general, finer grinds produces higher quality pellets since they can readily absorb moisture than large particles, and therefore, undergo a higher degree of conditioning. In addition, finer grinds have higher surface area of contact to form bonds/solid bridges during the compaction processes. Also, large particles are fissure points that cause cracks and fractures in compacts (MacBain, 1966). A reduction in hammer mill screen size from 3.2 to 0.6 mm can result in an increase in pellet densities from 5 to 16% (Kaliyan and Morey, 2006b; Kashaninejad et al., 2011; Mani et al., 2002 and 2004a). However, no significant trend in change in density were observed at geometric mean particles size of 0.6 mm and lower (Kaliyan and Morey 2006b; Mani et al., 2002). The change in pellet density depends on the type of biomass.

This chapter will address various factors that directly or indirectly effect densification of agricultural biomass residue into high quality pellets. The compression and compaction characteristics of ground biomass will be dealt in detail that will provide a comprehensive understanding of the behaviour of biomass as influenced by various factors. The compression studies will explore the affect of independent variables such as biomass, treatment, grind size, and moisture content on pellet density and durability, while compaction studies will study the effect of various machine variables on the pellet quality. In addition, overall specific energy requirements will be established and techno-economic models will be explained.

2. LIGNOCELLULOSIC COMPOSITION AND HIGHER HEATING VALUES

The experimental lignocellulosic composition of agricultural straw can be determined using the modified NREL LAP method for "Determination of Structural Carbohydrates and Lignin in Biomass" (Table 1) (Adapa et al., 2011; Sluiter et al., 2008). This procedure uses a two-step acid hydrolysis to fractionate the biomass into forms that are more easily quantified. During this process, the lignin fractionates into acid insoluble material and acid soluble material, while the polymeric carbohydrates are hydrolyzed into the monomeric forms, which are soluble in the hydrolysis liquid and subsequently can be measured using HPLC. The Percentage cellulose in the samples can be measured using the percentage glucan content, while the percentage hemicelluloses can be measured by adding the percentage mannose, galactose, xylose and arabinose content in the biomass samples.

Table 1 shows the lignocellulosic composition and higher heating values of non-treated and steam exploded barley, canola, oat and wheat straw samples. In general, the cellulose, hemicelluloses and lignin content of steam exploded straw was higher than non-treated straw. This may be due to other components (soluble lignin, loosely-bound sugars) being washed away during steam explosion, thereby leaving the proportion of insoluble lignin, cellulose and hemicellulose in the resulting dried sample higher than for the non-treated samples (i.e. higher percent of dry mass).

TABLE 1. Lignocellulosic composition and higher heating values of non-treated and steam exploded agricultural straw (Adapa et al., 2011)

Properties of Biomass	Barley Straw		Canola Straw		Oat Straw		Wheat Straw	
	NT	SE	NT	SE	NT	SE	NT	SE
Composition (% of dry matter)								
Cellulose[b]	22.7 ± 0.9[a]	25.3 ± 1.8	22.4 ± 0.8	27.5 ± 1.1	25.4 ± 1.0	27.4 ± 2.4	27.1 ± 1.0	29.9 ± 1.4
Hemicellulose[c]	21.2 ± 0.5	21.0 ± 1.4	16.9 ± 0.5	20.2 ± 0.7	21.7 ± 0.9	18.8 ± 1.2	21.1 ± 0.5	19.7 ± 0.9
Galactose	0.9 ± 0.0	0.7 ± 0.0	1.0 ± 0.0	0.9 ± 0.1	0.8 ± 0.0	0.7 ± 0.0	0.8 ± 0.0	0.9 ± 0.1
Mannose	1.6 ± 0.2	1.5 ± 0.0	2.3 ± 0.1	1.9 ± 0.4	1.4 ± 0.0	1.7 ± 0.1	1.6 ± 0.1	2.8 ± 0.2
Xylose	14.4 ± 0.3	15.3 ± 1.0	11.5 ± 0.5	14.3 ± 0.2	15.1 ± 0.8	13.3 ± 1.0	14.9 ± 0.4	13.5 ± 0.4
Arabinose	4.4 ± 0.2	3.5 ± 0.5	2.0 ± 0.1	3.2 ± 0.0	4.4 ± 0.2	3.1 ± 0.2	3.9 ± 0.1	2.6 ± 0.2
Total Lignin[d]	21.0 ± 0.6	21.6 ± 0.6	19.6 ± 0.6	22.3 ± 0.2	19.5 ± 0.6	23.7 ± 0.2	22.5 ± 0.7	24.2 ± 0.3
Soluble Lignin	1.6 ± 0.1	1.4 ± 0.1	1.6 ± 0.1	1.2 ± 0.1	1.5 ± 0.1	1.3 ± 0.1	1.4 ± 0.0	1.0 ± 0.1
Insoluble Lignin	19.4 ± 0.6	20.2 ± 0.6	18.0 ± 0.6	21.1 ± 0.1	17.9 ± 0.7	22.4 ± 0.1	21.0 ± 0.7	23.3 ± 0.4
Higher Heating Values (MJ/kg of dry matter)								
HHV (MJ/kg)	16.4±0.3‡†	17.4±0.1	16.7±0.3	18.3±0.0	16.4±0.1	17.8±0.0	17.0±0.2	17.8±0.0

DM - Dry Matter; NT - Non-Treated; SE - Steam Exploded; [a] Average and standard deviation of 3 replicates at 95% confidence interval; [b]%Cellulose = %glucan; [c]%Hemicellulose = %(mannose + galactose + xylose + arabinose);
[d]%Total Lignin = %(soluble lignin + insoluble lignin); HHV - Higher Heating Values (measured using Parr 1281 Bomb Calorimeter); ‡3 replicates; † 95% confidence interval

The calorific (heating) value of biomass feedstocks are indicative of the energy they possess as potential fuels. The gross calorific value (higher heating value, HHV) and the net calorific value (lower heating value, LHV) at constant pressure measures the enthalpy change of combustion with and without water condensed, respectively (Demirbaş, 2007). A bomb calorimeter can be used to determine the HHV of the non-treated and steam exploded straw in MJ/kg. In addition, the ASTM StandardD5865-03 (ASTM, 2003) test method for gross calorific value of coal and coke, can be used as a guideline for heating value testing (Table 1).

Cellulose, hemicelluloses and lignin are major components of a plant biomass. Therefore, a change in composition could potentially lead to change in HHV of the biomass (Adapa et al., 2010a). The Net combined percentage change of cellulose, hemicelluloses and lignin in steam exploded barley, canola, oat and wheat straw is 5%, 19%, 5% and 4% higher than non-treated straw, respectively. As a result, the average HHV of steam exploded barley, canola, oat and wheat straw was 6%, 10%, 9% and 5% higher than non-treated straw, respectively (Table 1).

3. LAB-SCALE PELLETING OF AGRICULTURAL BIOMASS

3.1. Compression Test

A compression apparatus having a close fit plunger die assembly can be used to make a single compact in one stroke of the plunger from ground straw samples (Adapa et al., 2006 and 2010a; Mani et al., 2004). The compression test should be performed to study the effect of independent variables such as biomass, treatment, grind size, and moisture content on pellet density and durability. In order to simulate frictional heating during commercial pelleting operation, the compression die should be maintained at pre-heat temperatures of 75 to 100°C (Adapa et al., 2006; Kaliyan and Morey, 2009;Mani et al., 2006). Different levels of pre-set compressive forces can be applied using the Instron testing machine. Typical pre-set loads in the range of 31.0 to 150.0 MPa are applied to make pellets.Figure 1 represents the photographs of pellets made from barley, canola, oat and wheat straw grinds from hammer mill screen sizes of 3.2, 1.6 and 0.8 mm (Adapa et al., 2010a).

3.2. Single-Pellet Density

The density of pellet is calculated from the mass and volume (measuring the length and diameter) of compacts. In general, the density of pellets from agricultural straw significantly increases with an increase in applied pressure at any specific hammer mill screen size. An increase in pressure results in plastic deformation of ground particles and consequently leads to pellets that have densities closer to their respective particle densities (Adapa et al., 2010a; Kaliyan and Morey, 2009; Mani et al., 2004). The Application of pre-treatment has been observed to significantly increase the pellet density since pre-treated straw has lower geometric particle diameters and significantly higher particle densities (Adapa et al., 2010a; Kashaninejad and Tabil, 2011). Usually, it has been reported that an increase in moisture content from 10% and up results in a significant decrease in pellet quality (Hill and Pulkinen, 1988; Li and Liu, 2000; Obernberger and Thek, 2004; Shaw and Tabil, 2007). In general, a decrease in hammer mill screen size results in an increase in pellet density (Adapa et al., 2010a; Kaliyan and Morey, 2009; Kashaninejad et al., 2011; Mani et al., 2004). A comprehensive literature on various single-pellet compression test data is provided in Table 2.

Adapa et al. (2010a) reported that the type of agricultural biomass did not have any significant effect on pellet density, while steam explosion pre-treatment, applied pressure and screen size had significant effects. In addition, correlation for pellet density with applied pressure and hammer mill screen size having highest R^2 values were developed (Table 3). Similarly, Kaliyan and Morey (2009) indicated that the pellet density of corn stover or switchgrass briquettes was significantly affected by pressure, particle size, moisture content and preheating temperature. Kashaninejad and Tabil (2011) also indicated that the pellets made from microwave-chemical pretreated biomass grinds had a significantly higher density and tensile strength than the untreated or samples pretreated by microwave-distilled water.

The densities of pellets should also be measured after a storage period of one week to one month to ascertain its dimensional stability, and associated handling and storage costs (Adapa et al., 2010b;Kaliyan and Morey, 2009). Adapa et al. (2010b) reported that a reduction in pellet density is usually expected due to relaxation of grinds in the pellet after release of pressure. They have observed that the relaxation was higher for larger hammer mill screen sizes and lower applied pressures. In some cases, the average reduction in density was negative giving the impression that pellet density actually increased during storage period. However, these negative values are primarily due to higher standard deviations in pellet density measurements. Therefore, from a practical manufacturing point of view, these values should be considered as a zero percent change in pellet density (Adapa et al., 2010b).

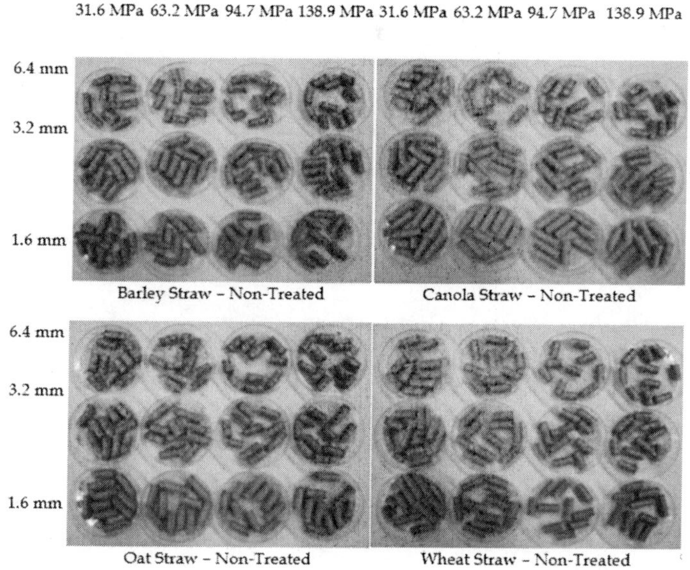

FIGURE 1. Photograph of pellets made from barley, canola, oat and wheat straw grind from hammer mill screen sizes of 3.2, 1.6, and 0.8 mm.

TABLE 2. Comprehensive literature review on single-pellet compression tests for agricultural biomass as feedstock for biofuel

Biomass	Independent Variables	Maximum Pellet Density	Reference
Barley, Canola, Oat and Wheat Straw	Hammer Mill Screen Size: 3.2, 1.6 and 0.8 mm Applied Pressure: 31.6, 63.2, 94.7 and 138.9 MPa Moisture Content: 10% (w.b.) Treatment: Non-Treated (NT) and Steam Exploded (SE)	Barley-NT 1003 kg/m^3 Barley-SE 1169 kg/m^3 Canola-NT 1035 kg/m^3 Canola-SE 1165 kg/m^3 Oat-NT 1024 kg/m^3 Oat-SE 1165 kg/m^3 Wheat-NT 1009 kg/m^3 Wheat-SE 1180 kg/m^3	Adapa et al., 2010a
Corn Stover and Switchgrass	Hammer Mill Screen Size: 3.0 and 4.6 mm Applied Pressure: 100 and 150 MPa Pre-Heat Temperature: 25, 75 and 100°C Moisture Content: 10 and 15% (w.b.)	Corn Stover 1197 kg/m^3 Switchgrass 1098 kg/m^3	Kaliyan and Morey, 2009
Barley and wheat straw	Hammer Mill Screen Size: 1.6 mm Applied Pressure: 126.3 MPa Moisture Content: 12% (w.b.) Treatment: Non-Treated (NT), Microwave Pretreated (MT) and Microwave-Chemical Pretreated (MCT)	Barley-NT 995 kg/m^3 Barley-MT 984 kg/m^3 Barley-MCT 1440 kg/m^3 Wheat -NT 950 kg/m^3 Wheat -MT 1032 kg/m^3 Wheat -MCT 1431 kg/m^3	Kashaninejad and Tabil, 2011
Barley and Wheat Straw, Corn Stover, and Switchgrass	Hammer Mill Screen Size: 3.2, 1.6 and 0.8 mm Applied Pressure: 31.6, 63.2, 94.7, 126.3 and 138.9 MPa Moisture Content: 6.22 to 8.30% (w.b.)	Barley Straw 1245 kg/m^3 Wheat Straw 1344 kg/m^3 Corn Stover 1399 kg/m^3 Switchgrass 1173 kg/m^3	Mani et al., 2004
Poplar Wood and Wheat Straw	Hammer Mill Screen Size: 3.2 and 0.8 mm Applied Pressure: 31.6, 63.2, 94.7 and 126.3 MPa Moisture Content: 9 and 15% (w.b.) Pre-Heat Die Temperature: 70 and 100°C Treatment: Non-Treated (NT) and Steam Exploded (SE)	Poplar-NT 1100 kg/m^3 Poplar-SE 1341 kg/m^3 Wheat-NT 1005 kg/m^3 Wheat-SE 1324 kg/m^3	Shaw, 2008

TABLE 3. Correlation for pellet density (ρ, kg/m³) with applied pressure (P, MPa) and hammer mill screen size (S, mm) for non-treated and steam exploded straw grinds.

Independent Variables and Interactions	Estimated Coefficients	R^2 value	Coefficient of Variation	Root Mean Square Error
$\rho_{BarleyNT} = 587.19 + 6.29(P) - 0.025(P)^2$				
Intercept	587.19	0.91	2.99	27.59
P	6.29			
P*P	-0.025			
$\rho_{BarleySE} = 727.38 + 7.41(P) - 0.03(P)^2 - 0.08(P \times S) + 2.84(S)^2$				
Intercept	727.38	0.89	3.30	34.89
P	7.41			
P*P	-0.03			
P*S	-0.08			
S*S	2.84			
$\rho_{CanolaNT} = 545.50 + 7.41(P) - 0.03(P)^2 - 0.11(P \times S)$				
Intercept	545.50	0.89	3.49	32.62
P	7.41			
P*P	-0.03			
P*S	-0.11			
$\rho_{CanolaSE} = 738.77 + 7.54(P) - 0.03(P)^2 + 0.11(P \times S) - 28.84(S)$				
Intercept	738.77	0.92	3.20	22.55
P	7.54			
P*P	-0.03			
P*S	0.11			
S	-28.84			
$\rho_{OatNT} = 666.24 + 5.79(P) - 0.02(P)^2 - 0.12(P \times S) + 3.54(S)^2$				
Intercept	666.24	0.87	3.13	29.00
P	5.79			
P*P	-0.02			
P*S	-0.12			
S*S	3.54			
$\rho_{OatSE} = 818.41 + 6.56(P) - 0.03(P)^2 + 0.12(P \times S) - 32.48(S)$				
Intercept	818.41	0.92	2.66	28.37
P	6.56			
P*P	-0.03			
P*S	0.12			
S	-32.48			
$\rho_{WheatNT} = 700.76 + 5.99(P) - 0.02(P)^2 - 36.06(S) + 3.48(S)^2$				
Intercept	700.76	0.90	3.01	27.91
P	5.99			
P*P	-0.02			
S	-36.06			
S*S	3.48			
$\rho_{WheatSE} = 717.60 + 6.77(P) - 0.03(P)^2$				
Intercept	717.60	0.91	2.95	31.55
P	6.77			
P*P	-0.03			

3.3. Durability

Durability represents the measure of shear and impact forces that a pellet could withstand during handling, storing and transportation process. The durability of pellets is usually measured following the ASABE Standard S269 (ASABE, 2007), which require about 50-100 g of pellets/ compacts. However, due to the limited number of pellets obtained during single-pellet compression test, it is not feasible to use this method. Instead, the durability of pellets can be measured by following the drop test method (Al-Widyan and Al-Jalil, 2001; Khankari et al., 1989; Sah et al., 1980; Shrivastava et al., 1989), where a single pellet is dropped from a 1.85 m height on a metal plate. The larger intact portion of the mass retained is expressed as the percentage of the initial weight.

Adapa et al. (2010) reported that the type of agricultural biomass, steam explosion pretreatment, applied pressure and screen size all had significant effect on pellet durability. Statistically, no significant correlation (R^2 values) was obtained for change in durability with applied pressure and hammer mill screen sizes. In general, pellet durability increases with an increase in applied pressure and grind size, and application of pre-treatment. Similarly, Kaliyan and Morey (2009) indicated that the durability of corn stover or switchgrass briquettes was significantly affected by pressure, moisture content and preheating temperature, while particle size did not have any significant effect.Kashaninejad et al. (2011) also reported the mean durability of pellets made of giant wild rye and mixed forage increased from 63.08 to 89.26% and from 61.47 to 89.21%, respectively when the hammer mill screen size increased from 0.8 to 3.2 mm. This could be primarily due to mechanical interlocking of relatively long fibers at higher grind sizes. They also indicated that at any specific compressive load, the pellet durability of biomass grinds with 12% moisture content was significantly higher than samples with 9 and 15% and demonstrates the moisture contents above or below 12% would lead to lower quality pellets.

3.4. Specific Energy for Compaction and Extrusion of Pellet

During the compression and extrusion processes of individual biomass compacts, the force-displacement data is recorded and can be used to calculate the specific compression and extrusion energies following the methodology reported by Adapa et al. (2006) and Mani et al. (2006). The area under the force-displacement curve can be integrated using the trapezoid rule (Cheney and Kincaid, 1980); when combined with the pellet mass, the specific energy values in MJ/t can be calculated.

During single-pellet compression and extrusion, the pellets arc prepared by densifying material against a base plate (representing the specific energy required to overcome friction within the straw grinds) as opposed to commercial operation where compacts are formed due to back-pressure effect in the die. Therefore, the specific energy required to extrude the compact should be included, which will closely emulate the specific energy required to overcome the friction between the ground compressed biomass and the die. Mani et al. (2006) have indicated that the extrusion (frictional) energy required to overcome the skin friction was roughly half of the total energy (12-30 MJ/t) for corn stover. Mewes (1959) showed that roughly 40% of the total applied energy was used to compress the materials (straw and hay) and the remaining 60% was used to overcome friction. Faborode and O'Callaghan (1987) studied the energy requirement for compression of fibrous agricultural materials. They reported that chopped barley straw at 8.3% (wb) moisture content consumed 28-31 MJ/t of energy, while un-chopped material consumed 18-27 MJ/t. Shaw (2008) reported that between 95 and 99% of the total specific energy was required to compress the grinds, whereas between 1 and 5% of the total specific energy was required to extrude the compact in single pellet tests. Shaw (2008) also reported that the mean values of specific compression energy ranged from 7.2 (pretreated wheat straw using steam explosion) to 39.1 MJ/t (wheat straw). Kashaninejad and Tabil (2011) indicated that microwave-distilled water and microwave-NaOH pre-treatments significantly increased the specific energy required for compression of wheat straw grinds so that it increased from 16.60 MJ/t to as high as 29.04 and 27.84 MJ/t after pre-treatment by microwave-distilled water and microwave-NaOH, respectively. They also reported less specific energy was required to compress wheat straw pre-treated by combination of microwave and $Ca(OH)_2$. More specific energy was required to eject the pre-treated wheat straw grinds than the untreated wheat straw grinds and it increased from 3.20 MJ/t to 23.08 after pre-treatment by microwave-NaOH. Data analysis showed that the total energy required for compression and ejection of wheat straw grinds pre-treated by microwave-distilled water or microwave-alkaline was higher than untreated samples.

Adapa et al. (2010b) reported that the type of agricultural biomass, steam explosion pretreatment, applied pressure and screen size all had significant effect on specific energy required to form a pellet. In addition, they have developed correlations for specific energy with applied pressure and hammer mill screen size having highest R^2 values for barley, canola, oat and wheat straw (Table 4). In general, the total and compression specific energy for compaction of non-treated and steam exploded barley, canola, oat and wheat straw at any particular hammer mill screen size significantly increased with an increase in applied pressure and significantly decreased with a decrease in hammer mill screen size.

Adapa et al. (2010b) also reported that the specific energy values obtained from the single-pellet compression tests should be used to compare the densification variables. However, these values may not have practical applications since the energy consumed by commercial densification machines / pilot-scale pellet mills may be higher.

4. COMPRESSION CHARACTERISTICS OF BIOMASS

4.1. Compression Mechanism

The compression characteristics of ground agricultural biomass vary under various applied pressures. It is important to understand the fundamental mechanism of the biomass compression process, which is required in the design of energy efficient compaction equipment to mitigate the cost of production and enhance the quality of the product (Mani et al., 2004). To a great extent, the strength of manufactured pellets depends on the physical forces that bond the particles together (Tabil and Sokhansanj, 1996). These physical forces come in three different forms during pelleting operations:

1. thermal;
2. mechanical; and
3. atomic forces (Adapa et al., 2002).

Pellets are formed by subjecting the biomass grinds to high pressures, wherein the particles are forced to agglomerate. It is generally accepted that the compression process is categorized in several distinct stages and difficult to let one simple monovariate equation to cover the entire densification region (Sonnergaard, 2001). Compression of grinds is usually achieved in three stages (Holman, 1991). In the first stage, particles rearrange themselves under low pressure to form close packing. The particles retain most of their original properties, although energy is dissipated due to inter-particle and particle-to-wall friction. During the second stage, elastic and plastic deformation of particles occurs, allowing them to flow into smaller void spaces, thus increasing inter-particle surface contact area and as a result, bonding forces like van der Waal forces become effective (Rumpf, 1962; Sastry and Fuerstenau, 1973;Pietsch, 1997). Brittle particles may fracture under stress, leading to mechanical interlocking (Gray, 1968). Finally, under high pressure the second stage of compression continues until the particle density of grinds has been reached. During this phase, the particles may reach their melting point and form very strong solid bridges upon cooling (Ghebre-Sellassie, 1989). Figure 2 shows the deformation mechanisms of ground particles under compression (Comoglu, 2007; Denny, 2002).

TABLE 4. Correlation for pellet density (ρ, kg/m³) with applied pressure (P, MPa) and hammer mill screen size (S, mm) for non-treated and stem exploded straw grinds.

Independent Variables and Interactions	Estimated Coefficients	R^2 value	Coefficient of Variation	Root Mean Square Error
$\rho_{BarleyNT} = 587.19 + 6.29(P) - 0.025(P)^2$				
Intercept	587.19	0.91	2.99	27.59
P	6.29			
P*P	-0.025			
$\rho_{BarleySE} = 727.38 + 7.41(P) - 0.03(P)^2 - 0.08(P \times S) + 2.84(S)^2$				
Intercept	727.38	0.89	3.30	34.89
P	7.41			
P*P	-0.03			
P*S	-0.08			
S*S	2.84			
$\rho_{CanolaNT} = 545.50 + 7.41(P) - 0.03(P)^2 - 0.11(P \times S)$				
Intercept	545.50	0.89	3.49	32.62
P	7.41			
P*P	-0.03			
P*S	-0.11			
$\rho_{CanolaSE} = 738.77 + 7.54(P) - 0.03(P)^2 + 0.11(P \times S) - 28.84(S)$				
Intercept	738.77	0.92	3.20	22.55
P	7.54			
P*P	-0.03			
P*S	0.11			
S	-28.84			
$\rho_{OatNT} = 666.24 + 5.79(P) - 0.02(P)^2 - 0.12(P \times S) + 3.54(S)^2$				
Intercept	666.24	0.87	3.13	29.00
P	5.79			
P*P	-0.02			
P*S	-0.12			
S*S	3.54			
$\rho_{OatSE} = 818.41 + 6.56(P) - 0.03(P)^2 + 0.12(P \times S) - 32.48(S)$				
Intercept	818.41	0.92	2.66	28.37
P	6.56			
P*P	-0.03			
P*S	0.12			
S	-32.48			
$\rho_{WheatNT} = 700.76 + 5.99(P) - 0.02(P)^2 - 36.06(S) + 3.48(S)^2$				
Intercept	700.76	0.90	3.01	27.91
P	5.99			
P*P	-0.02			
S	-36.06			
S*S	3.48			
$\rho_{WheatSE} = 717.60 + 6.77(P) - 0.03(P)^2$				
Intercept	717.60	0.91	2.95	31.55
P	6.77			
P*P	-0.03			

Note: ρ - Density, kg/m³; NT - Non-Treated; SE - Steam Exploded; P - Pressure, MPa; S - Hammer Mill Screen Size, mm

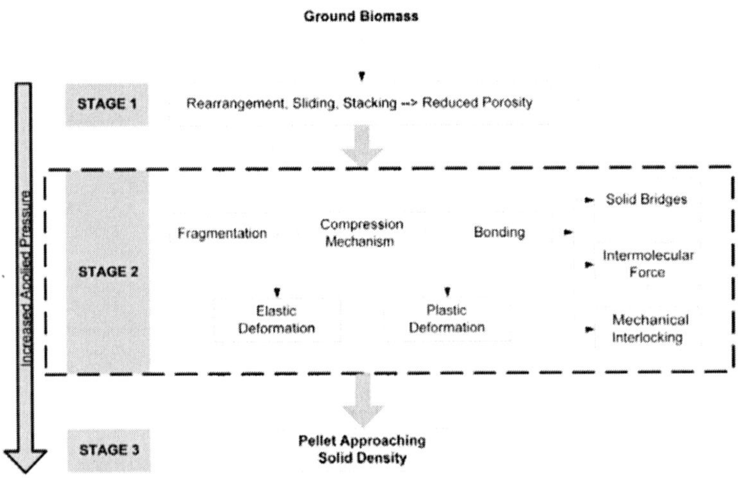

FIGURE 2. The deformation mechanisms of ground particles under compression (Comoglu, 2007; Denny, 2002)

4.2. Compression Characteristics Models

Densification or compaction of agricultural biomass grinds into pellets is an essential process towards production of biofuels. Ground biomass particles behave differently under different applied pressures (Adapa et al., 2002 and 2009a). Therefore, it is important to investigate the change in compact density and volume with pressures. One of the main purposes of fitting experimental data to an equation is usually to develop linear plots in order to make comparisons easier between different sets of data (Comoglu, 2007). A majority of compression models applied to biomass materials have been discussed and reviewed in detail by Adapa et al. (2002 and 2009a), Denny (2002) and Mani et al. (2003). Adapa et al. (2009a) reported that Kawakita and Ludde (1971), Cooper and Eaton (1962) and Jones (1960) models provided the best compression and deformation characteristics of agricultural biomass.

4.2.1. Jones model

Jones (1960) expressed the density-pressure data of compacted powder in the form of equation 1.

$$ln(\rho) = mln(P) + b \quad (1)$$

where, ρ is bulk density of compact powder mixture, kg/m^3, P is applied compressive pressure, MPa;m and b are model constants.

The constants b and m are determined from the intercept and slope, respectively, of the extrapolated linear region of the plot of $ln(\rho)$ vs $ln(P)$. The constant m has been shown to be equal to the reciprocal of the mean yield pressure required to induce plastic deformation (York and Pilpel, 1973). A large $m$value (low yield pressure) indicates the onset of plastic deformation at relatively low pressure, thus, the material is more compressible.

4.2.2. Cooper-eaton model

Cooper and Eaton (1962) studied the compaction behavior of four ceramic powders. In each case it, was assumed that compression is attained by two nearly independent probabilistic processes, namely, the filling of voids having equal size as particles and filling of voids smaller than particles. Based on these assumptions, the following equation (2) was given:

$$\frac{V_O - V}{V_O - V_S} = a_1 e^{-\frac{k_1}{P}} + a_2 e^{-\frac{k_2}{P}} \quad (2)$$

where, VO= volume of compact at zero pressure, m^3; V = volume of compact at pressure P, m^3; VS= void free solid material volume, m^3; $a\,1$, $a\,2$, $k\,1$, and $k\,2$ = Cooper-Eaton model constants.

The difficulty in practical use of equation (2) is the assignment of some physical significance to the constant parameters. In addition, another drawback of this model is its applicability to only one-component system (Comoglu, 2007).

4.2.3. Kawakita-ludde model

Kawakita and Ludde (1971) performed compression experiments and proposed an equation for compaction of powders based on observed relationship between pressure and volume (Equation 3).

$$\frac{P}{C} = \frac{1}{ab} + \frac{P}{a} \quad (3)$$

Where,

$$C = \frac{V_O - V}{V_O}$$

C = degree of volume reduction or engineering strain; a and b = Kawakita-Ludde model constants related to characteristic of the powder.

The linear relationship between P/C and P allows the constants to be evaluated graphically. This compression equation holds true for soft and low bulk density powders (Denny, 2002; Kawakita and Ludde, 1971), but particular attention must be paid on the measurement of the initial volume of the powder. Any deviations from this expression are sometimes due to fluctuations in the measured value of V_0. The constant a is equal to the values of $C = C\infty$ at infinitely large pressure P.

$$C_\infty = \frac{V_O - V_\infty}{V_O}$$

Where, V∝V∝= net volume of the powder, m^3.

It has been reported that the constant a is equal to the initial porosity of the sample, while constant $1/b$is related to the failure stress in the case of piston compression (Mani et al., 2004).

4.3. Compressibility of Different Biomass

The constant m in the Jones (1960) model can provided valuable information about the onset of plastic deformation of the ground agricultural biomass. It has been observed that ground particles obtained from larger hammer mill screen sizes has higher compressibility. In addition, application of pre-treatment also improves the compressibility of the agricultural biomass (Adapa et al., 2010).

The dimensionless coefficients, $a\,1$ and $a\,2$ in Cooper and Eaton (1962) model represent the densification of powdered material by particle rearrangement and deformation, respectively. If the sum of coefficients $(a\,1 + a\,2)$ is less than unity, it is an indication that other process must become operative before complete compaction is achieved. For agricultural biomass grinds, the $a\,1$ values were higher than $a\,2$ values, hence the material was primarily densified through the process of particle rearrangement. Occasionally, the sum of coefficients $(a\,1 + a\,2)$ for agricultural biomass was observed to be above unity. The phenomenon of having sum of coefficient more than unity was also observed by Adapa et al.

(2002 and 2009a), and Shivanand and Sprockel (1992), which implies that the densification could not be fully attributed to the two mechanisms of compression as assumed by theCooper and Eaton (1962) model (Adapa et al., 2010a).

In the Kawakita and Ludde (1971) model, constant *a* represents the initial porosity of the sample. It has been reported that the porosity and hammer mill screen sizes (corresponding geometric mean particle diameter) are positively correlated. In addition, porosity increases with application of pre-treatment since organized lignocellulosic structure of biomass disintegrates during this process. The parameter *1/b* in the Kawakita-Ludde model indicates the yield strength or failure stress of the compact. In general, the yield strength has negative correlation with hammer mill screen sizes. Also, application of pre-treatment lowers the yield strength of ground agricultural biomass. Statistically, theKawakita and Ludde (1971) model has been observed to provide accurate representation of the compression and deformation characteristics of agricultural biomass (Adapa et al., 2010a).

5. PILOT-SCALE PELLETING OF AGRICULTURAL BIOMASS

Pilot-scale densification of biomass is required to demonstrate the feasibility of production of pellets by application of various variables studied during single-pellet experiments. A pilot-scale pellet mill such as CPM CL-5 pellet mill (Figure 3) (California Pellet Mill Co., Crawfordsville, IN) can be used for processing of agricultural straw grinds into pellets. The pellet mill usually consists of a corrugated roller and ring die assembly, which compacts and extrudes the biomass grinds from the inside of a ring-shaped die by pressure applied by rolls where either the die or the roll suspension is rotating. Rolls are mounted close to the die surface, but still leaving room for a compacted feed layer to enter the roll gap. Friction between feed layer and rolls makes the rolls rotate (Larsson et al., 2008). In addition to variables indicated in the single-pellet testing, the quality of pellets also depends on machine variables such as the ring die size (radius), length (thickness, l), ring hole diameter (*d*), *l/d*ratio, and the rotational speed of the pellet mill (Adapa et al., 2002; Hill and Pulkinen, 1988; Tabil and Sokhansanj, 1996). A monitoring study of commercial pellets was done by Hill and Pulkinen (1988), on variables such as die geometry, conditioning temperatures, natural moisture of the grind, forage quality, bulk density of the grinds, and the use of binding agents. Similarly, Larsson et al. (2008) studied the effect of raw material moisture content, steam addition, raw material bulk density, and die temperature on production of high quality pellets. Also, Serrano et al. (2011), determined the effect of grind size, moisture content and customization of barley straw by adding pine dust to the mixture (blended pellets).

The feed rate of ground biomass to the pellet mill can be controlled using a vibratory feeder (Figure 3). The feed rate should be optimized according to the pellet mill capacity, which will directly affect the throughput. The pilot-scale pelleting test should be performed for a predefined period and the manufactured pellets should be collected and weighed to determine the pellet mill throughput (kg/h). In addition, the pellet mill energy consumption (kWh) should be recorded in real time using a data logger connected to a computer and should be used to calculate the specific energy (MJ/t) required to manufacture pellets from ground agricultural biomass.

Raw materials causing uneven pellet production have low bulk density compared to other milled biofuel pellet raw materials. Low raw material bulk density will put higher demands on the die feeding system of the pelletizer with greater volume throughput for maintained production level. Larsson et al. (2008) investigated the pre-compaction of reed canary grass as an alternative to avoid low and intermittent production of biofuel pellets. They have observed that the process of pre-compaction can increase the bulk density of raw material from 150 kg/m^3 to 270 kg/m^3, which resulted in the continuous production of pellets at a moisture content of 13.8% (w.b.). Pressurized steam conditioners are used in the feed pellet industry to decrease raw material porosity and to improve pellet hardness/ durability (Thomas et al., 1997). Adapa et al. (2010b) were unable to produce any pellets due to the low bulk density of both non-treated and pre-treated agricultural straw grinds at 10% moisture content (w.b.). Therefore, they have added moisture and oil to increase bulk density of grinds to a level of 17.5% (w.b.) and 10% (by weight), respectively, which resulted in production of pellets. Similar observation was made by Serrano et al. (2011) where they have to increase the grind moisture content in the range of 19-23% (w.b.) to produce pellets in a pellet mill. However, addition of pine sawdust to barley straw resulted in high quality pellets at a lower moisture content of 12% (w.b.).

Testing and, if required improving the durability of pellets is important for the industry to evaluate pellet quality and minimize losses during handling and transportation. The concept is not to add any external binders to enhance pellet quality, but rather activate the natural binders in the agricultural biomass by application of various variables, pre-processing techniques and pre-treatments. Biomass pellets can be customized based on proximate analysis data to make them suitable for direct combustion and thermo-chemical conversion applications. Customization can be achieved by forming composites of different straws to control important variables such as energy and ash content of pellets. Similarly, addition of biomass having good binding characteristics to straw with less cohesive characteristics may enhance particle bonding resulting in durable pellets.

Adapa et al. (2010b) reported pellet mill tests on both non-treated and steam exploded agricultural biomass at different hammer mill screen sizes. They have successfully produced pellets from ground non-treated barley, canola, oat and wheat straw at hammer mill screen sizes of 0.8 and 1.6 mm having moisture content of 17.5% (wb) and flax seed oil of 10% by weight. The non-treated ground straw at 3.2 and 6.4 mm screen size did not produce pellets. Similar pelleting process was followed for ground steam exploded straw. Due to very low bulk density and poor flowability, the steam exploded grinds did not produce pellets at any of the hammer mill screen sizes used in the investigation. However, the customized barley, canola, oat and wheat straw having 25% steam exploded material by weight at 0.8 mm screen size successfully produced pellets. Addition of higher percentage of steam exploded straw and customization at screen sizes of 1.6, 3.2, and 6.4 mm did not produce pellets, which could be due to the fact that adding steam exploded (having very low bulk density) to non-treated straw (having relatively higher bulk density) decreased the overall bulk density and flowability of the grinds, thus hindering the production of pellets in the pilot scale mill. The pilot scale pellet mill in this test is constrained with a small motor (3.7 kW (5 hp)) running it, whereas in a commercial pellet mill, the motors are much bigger and more tolerant to changes in feed bulk density. Shaw et al. (2007) reported similar trends where the quality of wheat straw pellets increased with an increase in moisture content to 15.9% (wb).

Figure 4 shows the photograph of pellets manufactured from barley, canola, oat and wheat straw from non-treated grinds at 0.8 and 1.6 mm screen sizes, and customized straw grinds at 0.8 mm having 25% steam exploded straw by weight (Adapa et al., 2010b).

5.1. Pellet Bulk Density

The mass, length and diameter of individual pellets should be used to determine individual pellet density in kg/m^3. The bulk density of manufactured pellets can be calculated by measuring the mass of pellets filled in a cylindrical container of known volume.

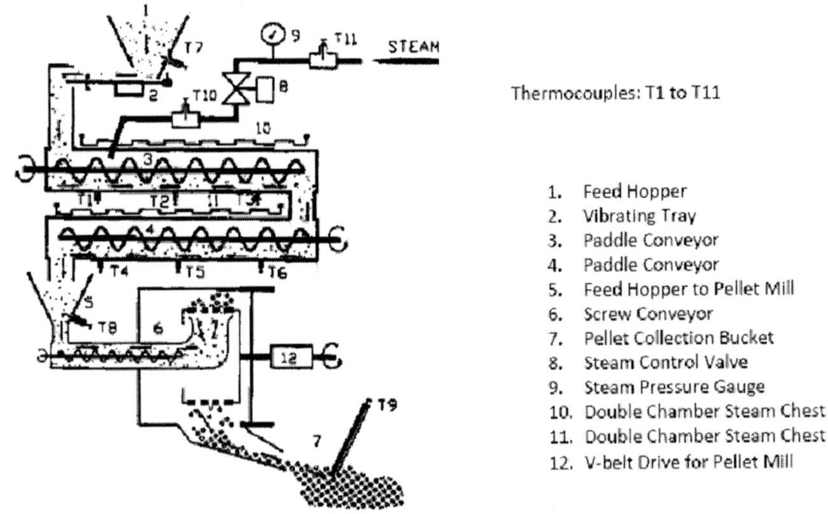

FIGURE 3. Schematic diagram of CPM CL-5 pellet mill

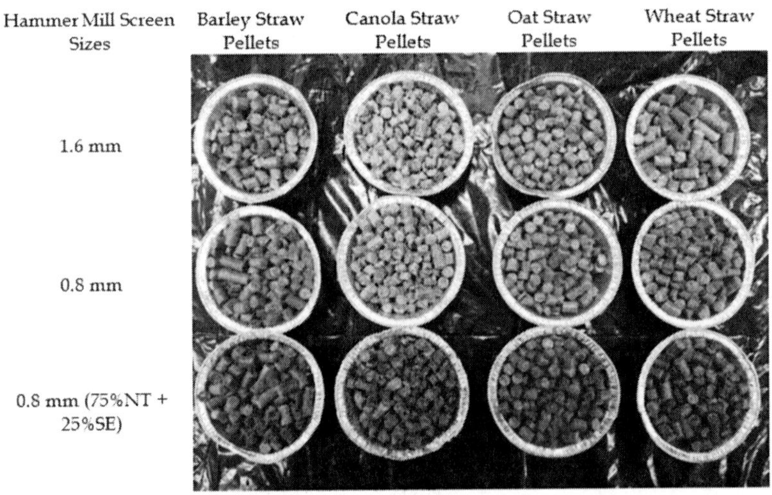

FIGURE 4. Photograph of pellets manufactured using a pilot scale pellet mill for non-treated (NT) straw at 1.6 and 0.8 mm hammer mill screen size, and customized grinds at 0.8 mm screen size having 25% steam exploded (SE) straw.

Larsson et al. (2008) reported that the most influential factor for the pellet bulk density was raw material moisture content and showing a negative correlation. Similarly, two other studies have observed that the bulk density of wheat straw, big bluestem grass, corn stover, sorghum stalk and switchgrass decreased with an increase in moisture content (Colley et al., 2006; Theerarattananoon, et al., 2011). Larsson et al. (2008) did not find any correlation between pellet bulk density and die temperature, which contradicts to the observations made by Hill and Pulkinen,1988, and Tabil and Sokhansanj (1996). Serrano et al. (2011) did not observe any significant effect of hammer mill screen size (4 mm and 7 mm) on pellet density. However, pellet density decreased with an increase in moisture content.

Adapa et al. (2010b) reported pellet density obtained from non-treated straw samples at 1.6 and 0.8 mm, and customized sample having 25% steam exploded straw at 0.8 mm screen size (Table 5). In general, pellet density increased with a decrease in screen size from 1.6 to 0.8 mm. However, no significant differences in density values were observed for non-treated samples at 0.8 mm and customized samples, except for canola and oat straw. This could be due to large fluctuation in individual pellet density values. All of the pellet density values reached near individual biomass particle densities at respective grind sizes (Adapa et al., 2010b).

Bulk density of pellets from barley, canola, oat and wheat straw showed significant difference with grind size and customization, except for wheat straw pellets at 0.8 mm for non-treated and customized samples (Table 5). In general, the average pellet bulk densities obtained for customized straw samples were higher (except for barley straw), which is consistent with increase in particle densities. The bulk densities of pellets manufactured were higher than the minimum design value of 650 kg/m^3 suggested by Obernberger and Thek (2004) for wood pellet producers.

5.2. Pellet Durability

The durability of pellets can be measured following the ASABE Standard S269 (2007). The method states that 100 g of pellet sample should be weighed and placed in a dust-tight enclosure/ chamber, and tumbled for 10 min at 50 r/min. A 5.70 mm sieve should used to determine the fines produced by the pellets during the tumbling process. The mass of pellets left on the sieve, as percentage of the total mass of pellet sample used during the test, was considered as the durability of the pellets.

Hill and Pulkinen (1988) reported an increase in pellet durability by about 30 to 35% with an increase in pellet temperature from 60 to 104ºC. A die length-to-diameter ratio (*l/d*) of 8 to 10 was also reported to be ideal for making high quality pellets. Similarly, Tabil and Sokhansanj (1996) conducted a study for improving the physical quality of alfalfa pellets by controlling and optimizing the manufacturing process. The process conditions investigated were steam conditioning temperatures, die geometry (length to diameter or *l/d* ratio), hammer mill screen sizes used in grinding dry chops, and die speed. They reported that a higher conditioning temperature (95ºC) resulted in improved durability of processed pellets. The durability of samples was generally better using the smaller die (higher *l/d*ratio). The hammer mill screen size did not show any affect on pellet durability. Finally, they reported that high durable pellets are obtained at low die speed (250 rpm). Theerarattananoon et al. (2011) also observed that an increase in length / thickness of die resulted in significant increase in durability of biomass pellets.

Larsson et al. (2008) and Serrano et al. (2011) reported that the most influential factor affecting pellet durability was raw material moisture content and showing a positive correlation. The maximum durability was obtained at moisture content of 14.9% without steam addition, and at 13.7% (w.b.) with steam addition of 2.6% (Larsson et al., 2008). Serrano et al. (2011) found that the highest mechanical durability reached for barley straw pellets was 95.5% at moisture content of 19-23% (w.b.), while no pellets were formed below the 19% moisture content. In addition, they observed that the durability of barley straw pellets increased with addition of pine sawdust at 2, 7 and 12% bu weight. However, the percentage increase in pine sawdust did not have significant effect on durability. Colley et al. (2006) observed highest durability of 95.9% for pellets from switchgrass at a moisture content of 8.6% (w.b.). The moisture content in the range of 9-14%, 9-11% and 14-16% (d.b.) did not have any significant effect on maximum durability of 96.8%, 96.8% and 89.5% for pellets from wheat straw and corn stover, big bluestem grass, and sorghum stalk, respectively (Theerarattananoon, et al., 2011); however, further increasing the moisture content reduced pellet durability for respective agricultural biomass. Also, the durability was observed to be positively correlated to die temperature (Larsson et al., 2008).

Adapa et al. (2010b) reported that the durability of pellets obtained from non-treated straw samples at 1.6 and 0.8 mm, and customized sample having 25% steam exploded straw at 0.8 mm screen size were significantly different (Table 5). In general, higher durability values were observed for non-treated straw samples at 0.8 mm hammer mill screen size. The durability of pellets significantly increased with a decrease in grind size for non-treated samples from 1.6 to 0.8 mm. However, addition of steam exploded straw to non-treated straw at 0.8 mm screen size resulted in a decrease in durability, except for wheat straw. This could be due to the fact that steam exploded material has lower soluble lignin content and higher cellulose and hemicelluloses content compared to non-treated straw (Table 1). This observation is in contrast to Lam et al. (2008), who reported that the quality (durability) of pellets produced from steam exploded sawdust was 20% higher than non-treated sawdust. Though, it is important to note that high durability values (>80%) were obtained for all pilot scale pelleting tests.

5.3. Specific Energy and Energy Balance During Pelleting

The durability of pellets was negatively correlated to pellet mill throughput and was positively correlated to specific energy consumption (Table 5). The specific energy values obtained from pilot scale pellet mill are 10-25 times higher than reported by Mani et al. (2006b) and Adapa et al. (2010aand 2009b) for agricultural straw, using a single pellet Instron testing machine. The higher pellet mill specific energy numbers could be due to higher friction values and practical pelleting conditions, which are closer to industrial operations.

An overall specific energy analysis is desired in order to understand the net amount of energy available for the production of biofuels after postharvest processing and densification of agricultural straw. The specific energy analysis was performed for pilot-scale pelleting of non-treated and customized (75% non-treated + 25% steam exploded) barley, canola, oat and wheat straw at 1.6 and 0.8 mm hammer mill screen sizes (Table 6). The specific energy for grinding of straw at 0.8 mm was calculated using regression equations reported in Adapa et al. (2011b). The specific energy for chopping and grinding of biomass, production of pellets using pellet mill and higher heating values for straw were obtained from experimental data (Adapa et al., 2011b and 2010b). In addition, the specific energy required for operating the chopper, hammer mill and pellet mill were 337, 759 and 429 W, respectively. On average, the operation of biomass chopper required five times more energy than chopping of biomass. On the other hand, the grinding of biomass required on an average three times more energy than operation of hammer mill. Interestingly, almost the same amount of energy was required to operate the pellet mill and production of pellets. Total specific energy required to form pellets increased with a decrease in hammer mill screen size from 1.6 to 0.8 mm, however, the total specific energy for the process decreased for customized straw compared to non-treated straw at 0.8 mm screen size (Table 6). It has been determined that the net specific energy available for production of biofuel is a significant portion of original agricultural biomass energy (92-94%) for all agricultural biomass (Table 6).

TABLE 5. Pellet density, durability, throughput and specific energy data for non-treated and steam exploded barley canola, oat and wheat straw at 17.5% moisture content (wb) and 10% flaxseed oil content

Agricultural Biomass	Hammer Mill Screen Size (mm)	Pellet Density (kg/m³)	Pellet Bulk Density (kg/m³)	Durability (%)	Throughput (kg/h)	Specific Energy (MJ/t)
Barley Straw	1.6 (100% NT)	1158±109*†£ aD	665±01‡ aD	91±00‡ aD	4.88	293
	0.8 (100% NT)	1174±46 aD	700±07 bD	93±01 bD	4.21	353
	0.8 (75% NT + 25% SE)	1184±63 aD	714±02 cD	87±01 cD	3.46	301
Canola Straw	1.6 (100% NT)	1023±85 aE	629±01 aE	90±01 aD	3.86	385
	0.8 (100% NT)	1204±43 bDE	720±04 bE	95±00 bE	3.63	440
	0.8 (75% NT + 25% SE)	1144±50 cD	641±01 cE	82±00 cE	5.51	265
Oat Straw	1.6 (100% NT)	1140±63 abD	631±03 aE	89±01 aE	4.48	340
	0.8 (100% NT)	1188±78 aDE	649±02 bF	93±00 bD	3.81	344
	0.8 (75% NT + 25% SE)	1071±101 bE	676±06 cF	89±01 aF	4.03	335
Wheat Straw	1.6 (100% NT)	1163±57 aD	673±02 aF	94±01 aF	5.44	381
	0.8 (100% NT)	1278±136 bE	721±04 bE	95±01 bE	3.81	297
	0.8 (75% NT + 25% SE)	1213±88 abD	722±04 bG	95±00 cG	4.08	342

NT - Non-treated Straw Samples; SE - Steam Exploded Straw Samples; *10 replicates; ‡3 replicates; † 95% confidence interval; £ Student-Neuman-Keuls test at 5% level of significance for same sample biomass at various hammer mill screen sizes (a, b and c); at same hammer mill screen size for different sample biomass (D, E, F and G)

5.4. Cost and Life Cycle Assessment of Biomass Densification

Sultana et al. (2010) performed a techno-economic analysis and developed a model for a plant that can produce agricultural straw (barley, oat and wheat) pellets for 30 years. They have included the cost of obtaining the straw, transporting straw to the pellet plant, and producing pellets. Costs incurred by the plant for the production of pellets included capital cost, energy cost, labor cost, and consumable cost. The biomass procurement area was determined to estimate the transportation cost. The scale factors for all the equipment related to pellet production were determined based on the data of previous studies (Sultana et al., 2010). To develop the model, minimum, average and maximum yields of wheat, barley and oats straw in Western Canada were considered. They have determined that the cost of pellets does not change much for capacities over 70,000 tonnes per year (cost of production per tonne is $170.89). Therefore, the optimum size is the same for both average and maximum yield cases. In addition, it was observed that the total cost of pellet production is most sensitive to field cost followed by the transportation cost.

TABLE 6. Overall specific energy analysis to show net energy available for production of biofuels after postharvest processing and densification of agricultural straw.

Treatment	Hammer Mill Screen Size (mm)	Specific Energy (MJ/t)				HHV (MJ/t)	Net Energy [γ] (MJ/t)
		Chopping Biomass	Grinding Biomass	Pilot-Scale Pelleting	Total[£]		
Barley							
NT*	1.6	11.3	90.4	293	924	16400	15476
NT	0.8	11.3	206.6	353	1100	16400	15300
75% NT + 25% SE*	0.8	11.3	189.3	301	1030	16650	15620
Canola							
NT	1.6	7.1	128.5	385	987	16700	15713
NT	0.8	7.1	363.3	440	1277	16700	15423
75% NT + 25% SE	0.8	7.1	341.6	265	1080	17100	16020
Oat							
NT	1.6	9.9	149.5	340	1029	16400	15371
NT	0.8	9.9	253.6	344	1137	16400	15263
75% NT + 25% SE	0.8	9.9	245.2	335	1120	16750	15630
Wheat							
NT	1.6	8.2	153.3	381	1048	17000	15952
NT	0.8	8.2	382.7	297	1194	17000	15806
75% NT + 25% SE	0.8	8.2	332.1	342	1188	17200	16012

*NT - Non-Treated; SE - Steam Exploded
£ Total Specific Energy = Specific Energy (Chopping Biomass + Operating Chopper + Grinding Biomass + Operating Hammer Mill + Pilot-Scale Pelleting)
γNet Energy = HHV - Total

Life cycle assessment (LCA) study was performed on wheat straw production system and densification system in the Canadian Prairies using the LCA modelling software tool SimaPro 7.2 to determine the environmental burdens of manufacturing the wheat straw bale and wheat straw pellet (Li et al., 2011). The factors taken into consideration were greenhouse gas emission, acidification, eutrophication, ozone layer depletion, abiotic depletion, human toxicity, and photochemical oxidation. Li et al. (2011) reported that the production of biomass pellet has higher global warming effect than biomass bale, especially in CO_2 and CH_4 emissions from fossil fuel consumption, which is very high in densification system due to machinery usage. It was also reported that the production of wheat straw pellet has higher environmental impact on acidification, eutrophication, human toxicity and other categories than biomass bale. The dominant factors determining most environmental impacts in agricultural system are fertilizer use and production, while machinery use, manufacturing and energy consumption are main contributors to greenhouse gas emission and other environmental burdens in the densification system (Li et al., 2011).

6. SUMMARY

The densification of biomass into durable compacts is an effective solution to meet the requirement of raw material for biofuel production. The compression characteristics of ground agricultural biomass vary under various applied pressures. It is important to understand the fundamental mechanism of the biomass compression process, which is required to design an energy efficient compaction equipment to mitigate the cost of production and enhance the quality of the product. To a great extent, the strength of manufactured compacts depends on the physical forces that bond the particles together. These physical forces are generated in three different forms during compaction operations: a) thermal; b) mechanical; and c) atomic forces. To customize and manufacture high quality products that can withstand various forces during transportation and handling, it is essential to predict desirable and dependent quality parameters (density and durability) with respect to various independent variables (pre-treatment, grind size, applied pressure, hold time, die temperature, and moisture content). In addition, specific energy requirements of manufacturing biomass pellets should be established, which can assist in determining the economic viability of densification process.

The density of biomass pellet has been observed to significantly increase with an increase in applied pressure and a decrease in hammer mill screen size. In addition, application of pre-treatment has observed to significantly increase the pellet density since pre-treated straw has lower geometric particle diameters and significantly higher particle densities. Statistically, agricultural biomass did not have any significant effect on pellet density, while steam explosion pre-treatment, applied pressure, moisture content, pre-heat temperature and screen size had significant effect. A negative correlation has been observed between the pellet bulk density and moisture content, while a positive correlation exists between bulk density and pellet mill die temperature. In general, average pellet bulk densities obtained for customized straw samples is higher as a direct result of increase in particle densities.

Agricultural biomass, steam explosion pre-treatment, applied pressure, moisture content, pre-heat temperature and screen size all had significant effect on pellet durability. In general, durability of pellets increases with an increase in applied pressure and grind size, and application of pre-treatment. An increase in pellet mill die temperature, steam conditioning temperature and die thickness resulted in an increase in pellet durability. No specific trend in durability was observed with customization of straw by mixing non-treated and steam exploded straw grinds.

The specific energy required to form a pellet has been significantly affected by the type of agricultural biomass, steam explosion pre-treatment, applied pressure and screen size. The total and compression specific energy for compaction of non-treated and steam exploded barley, canola, oat and wheat straw at any particular hammer mill screen size significantly increased with an increase in applied pressure and significantly decreased with a decrease in hammer mill screen size. Durability of pellets was negatively correlated to pellet mill throughput and was positively correlated to specific energy consumption. An overall energy balance was performed, which showed that a significant portion of original agricultural biomass energy (92-94%) is available for the production of biofuels.

REFERENCES

1. P. K. Adapa, L. G. Tabil, G. J. Schoenau, T. Canam, T. Dumonceaux, 2011aQuantitative Analysis of Lignocellulosic Components of Non-Treated and Steam Exploded Barley, Canola, Oat and Wheat Straw using Fourier Transform Infrared Spectroscopy. The Journal of Agricultural Science and Technology, 5(7), (In-Press).
2. Adapa, P.K., Tabil, L.G., & Schoenau, G.J.2011bGrinding Performance and Physical Properties of Non-Treated and Steam Exploded Barley, Canola, Oat and Wheat Straw. Journal of Biomass and Bioenergy,doi:10.1016/j.biombioe.2010.10.004, 35,549 561
3. P. K. Adapa, L. G. Tabil, G. J. Schoenau, 2010a Compression Characteristics of Non-Treated and Steam Exploded Barley, Canola, Oat and Wheat Straw Grinds.ASABE Journal of Applied Engineering in Agriculture 617 632
4. P. K. Adapa, L. G. Tabil, G. J. Schoenau, 2010b Pelleting Characteristics of Selected Biomass with and without Steam Explosion Pretreatment International Journal of Agricultural and Biological Engineering 62 79
5. P. K. Adapa, L. G. Tabil, G. J. Schoenau, 2009aCompression Characteristics of Selected Ground Agricultural Biomass. Agricultural

Engineering International: the CIGR Ejournal, Manuscript 1347, XI(June), 1 19

6. Adapa, P.K., Tabil, L.G., & Schoenau, G.J.2009bCompaction Characteristics of Barley, Canola, Oat and Wheat Straw. Biosystems Engineeringdoi:10.1016/j.biosystemseng.2009.06.022,104,335 344
7. P. K. Adapa, G. J. Schoenau, L. G. Tabil, E. A. Arinze, A. Singh, A. K. Dalai, 2007Customized and value-added high quality alfalfa products- a new concept. Agricultural Engineering International: the CIGR Ejournal, Manuscript FP 07 003, IX(June), 1 28
8. P. K. Adapa, A. Singh, G. J. Schoenau, L. G. Tabil, 2006Pelleting Characteristics of Fractionated Alfalfa Grinds- Hardness Models. International Journal of Powder Handling and Processing, 18(5), 294 299
9. P. K. Adapa, L. G. Tabil, G. J. Schoenau, B. Crerar, S. Sokhansanj, 2002Compression Characteristics of Fractionated Alfalfa Grinds. Powder Handling and Processing, 14(4), 252 259
10. M. I. Al-Widyan, H. F. Al-Jalil, 2001Stress-density relationship and energy requirement of compressed only cake. Applied Engineering in Agriculture, 17(6), 749 753
11. ASABE Standards 269.4. 2007Cubes, Pellets and Crumbles- Definitions and methods for determining density, durability and moisture content. In: ASABE Standards. St. Joseph, MI.: American Society of Agricultural and Biological Engineers.
12. A. S. T. M. A. S. T. M. D5865-0, 2003Standard test method for gross calorific value of coal and coke. In: Annual book of ASTM standards, 05West Conshohocken, PA.: American Society for Testing and Materials.
13. J. L. Bowyer, V. E. Stockmann, 2001Agricultural residues: An exciting bio-based raw material for the global panel industry. Forest Products Journal, 51(1), 10 21
14. Campbell, C.A., Zentner, R.P., Gameda, S., Blomert, B. & Wall, D.D.2002Production of annual crops on the canadian prairies: trends during 1976- 1998. Canadian Journal of Soil Science, 8245 57
15. W. J. Chancellor, 1962Formation of hay wafers with impact loads. Agricultural Engineering, 43(3), 136 138
16. W. Cheney, D. Kincaid, 1980 Numerical mathematics and computingMonterey, CA: Brooks/Cole Publishing Company.

17. Colley, Z., Fasina, O.O., Bransby, D., & Lee, Y.Y. (2006). pp. 1845-1851 Moisture effect on the physical characteristics of switchgrass pellets. Transactions of ASABE, 49,1845 1851
18. T. Comoglu, 2007An Overview of Compaction Equations. Journal of Faculty of Pharmacy, Ankara 36(2), 123 133
19. A. R. Cooper, L. E. Eaton, 1962 Compaction Behavior of Several Ceramic Powders Journal of the American Ceramic Society 97 101
20. A. Demirbaş, 2007Effect of moisture and hydrogen content on the heating value of fuels. Energy Sources Part A, 29, 649 655
21. A. Demirbaş, A. Sahin, 1998Evaluation of biomass residue 1. Briquetting waste paper and wheat straw mixtures. Fuel Processing Technology, 55 175 183
22. P. J. Denny, 2002 Compaction Equations: A Comparison of the Heckel and Kawakita Equations Powder Technology 162 172
23. M. O. Faborode, J. R. O'Callaghan, 1987 Optimizing the compression/ briquetting of fibrous agricultural materials Journal of Agricultural Engineering Research 245 262
24. FAO, 2011Gasification Fuels. Food and Agriculture Organization of the United Nations. Website accessed on March 15, Web Address:http://www.fao.org/docrep/t0512e/t0512e0b.htm
25. I. Ghebre-Sellassie, 1989Mechanism of Pellet Formation and Growth. Pharmaceutical Pelletization Technology, ed. I. Ghebre-Sellassie, 123 143New York, NY: Marcel Dekker Inc.
26. W. A. Gray, 1968Compaction after Deposition. In The Packing of Solid Particles, 89 107New York, NY: Marcel Dekker Inc.
27. P. D. Grover, S. K. Mishra, 1996 Biomass Briquetting Technology and Practices.Food and Agriculture Organization (FAO), UN, Document 46
28. A. S. Gustafson, W. L. Kjelgaard, 1963 Hay pellet geometry and stability. AgriculturalEngineering, 44(8), 442 445
29. B. Hill, D. A. Pulkinen, 1988 A study of the factors affecting pellet durability and pelleting efficiency in the production of dehydrated alfalfa pellets. Saskatchewan, Canada: Saskatchewan Dehydrators Association.
30. C. A. Holley, 1983The densification of biomass by roll briquetting. Proceedings of the Institute for Briquetting and Agglomeration (IBA), 18, 95 102
31. L. E. Holman, 1991 The Compaction Behaviour of Particulate Materials: An Elucidation based on Percolation Theory Powder Technology 265

32. W. D. Jones, 1960Fundamental Principles of Powder Metallurgy; London UK; Edward Arnold Publishers Ltd.: 242 370
33. N. Kaliyan, R. V. Morey, 2006a Factors affecting strength and durability of densified products. ASABEAnnual International Meeting, American Society of Agricultural and Biological Engineers, Portland, Oregon July 9 12Paper 066077Niles Road, St. Joseph, MI 49085-9659 USA.
34. N. Kaliyan, R. V. Morey, 2006b Densification characteristics of corn stover and switchgrassASABE Annual International Meeting, American Society of Agricultural and Biological Engineers, Portland, Oregon July 9 12Paper 066174Niles Road, St. Joseph, MI 49085-9659 USA.
35. N. Kaliyan, R. V. Morey, 2009 Densification Characteristics of Corn Stover and Switchgrass Transactions of the ASABE 907 920
36. M. Kashaninejad, L. G. Tabil, 2011 Effect of microwave-chemical pre-treatment on compression characteristics of biomass grinds Biosystems Engineering 36 45
37. M. Kashaninejad, L. G. Tabil, X. Tabil, 2011Compression characteristics of giant wild rye and mixed forage grinds, Renewable Energy, Under review.
38. K. Kawakita, K. Ludde, H. , 1971 Some Considerations on Powder Compression Equations Powder Technology 61 68
39. K. K. Khankari, M. Shrivastava, R. V. Morey, 1989Densification characteristics of rice hulls. ASAE Paper 89-6093 89 6093St. Joseph, Mich., ASABE.
40. P. S. Lam, S. Sokhansanj, X. Bi, C. J. Lim, S. Mani, S. Melin, 2008Energetic feasibility of biomass densification with steam explosion pretreatment. CSBE Annual General Meeting; July 14 16Poster CSBE08305Vancouver, BC.
41. S. H. Larsson, M. Thyrel, P. Geladi, T. A. Lestander, 2008High quality biofuel pellet production from pre-compacted low density raw materials. Bioresource Technology, 99, 7176 7182
42. Y. Li, H. Liu, 2000 High-pressure densification of wood residues to form an upgraded fuel Biomass and Bioenergy 177 186
43. R. Liu, H. Yu, Y. Huang, 2005 Structure and morphology of cellulose in wheat straw Cellulose 25 34
44. Bain. R. Mac, 1966Pelleting animal feed. Chicago, IL: American Feed Manufacturing Association.

45. S. Mani, L. G. Tabil, S. Sokhansanj, 2006 Effects of compressive force, particle size and moisture content on mechanical properties of biomass pellets from grasses Biomass and Bioenergy 1420 1426
46. S. Mani, L. G. Tabil, S. Sokhansanj, 2006b Specific energy requirement for compacting corn stover Bioresource Technology 1420 1426
47. S. Mani, L. G. Tabil, S. Sokhansanj, 2004 Evaluation of compaction equations applied to four biomass species Canadian Biosystems Engineering 3
48. S. Mani, L. G. Tabil, S. Sokhansanj, 2003An overview of compaction of biomass grinds. Powder Handling & Process, 15(3),160 168
49. S. Mani, L. G. Tabil, S. Sokhansanj, 2002Compaction behavior of some biomass grinds. AIC Paper 02-305 02 305Saskatoon, Saskatchewan: AIC Meeting, CSAE/SCGR Program.
50. J. Mc Mullen, O. O. Fasina, C. W. Wood, Y. Feng, 2005 Storage and handling characteristics of pellets from poultry litter Applied Engineering in Agriculture 645 651
51. E. Mewes, 1959Verdichtungsgesetzmassigkeiten nach presstopfversuchen (Compression relationships as a result of experiments in pressure chambers). Landtechnische Forschung, 9(3), 68 76
52. I. Obernberger, G. Thek, 2004Physical characterization and chemical composition of densified biomass fuels with regard to their combustion behavior. Biomass and Bioenergy, 27, 653 669
53. W. Pietsch, 1997 Size Enlargement by AgglomerationIn Handbook of Powder Science and Technology, 2nd edition, eds. M.E. Fayed and L. Otten, 202 377Florence, KY: International Thomson Publishing.
54. A. Rajvanshi, 1986Published as Chapter 4 in the book Alternative Energy in Agriculture, Vol. II, Ed. D. Yogi Goswami, CRC Press, Boca Raton, FL.
55. H. Rumpf, 1962The Strength of Granules and Agglomerates. In Agglomeration, ed. W.A. Knepper, 379 419New York, NY: John Wiley and Sons.
56. P. Sah, B. Singh, U. Agrawal, 1980Compaction behavior of straw. Journal of Agricultural Engineering-India, 18(1), 89 96
57. K. V. S. Sastry, D. W. Fuerstenau, 1973 Mechanisms of Agglomerate Growth in Green Pelletization Powder Technology 97 105
58. C. Serrano, E. Monedero, M. Laupuerta, H. Portero, 2011 Effect of Moisture Content, Particle Size and Pine Addition on Quality

Parameters of Barley Straw Pellets Fuel Processing Technology 699 706

59. M. D. Shaw, L. G. Tabil, 2007 Compression and relaxation characteristics of selected biomass grinds.ASAE Annual International Meeting, Minneapolis, MN, June 17 20Paper 076183Niles Road, St. Joseph, MI 49085-9659 USA.
60. M. D. Shaw, L. G. Tabil, A. Opoku, 2007Pelleting characteristics and related properties of poplar wood and wheat straw grinds. A report prepared for SunOpta Bioprocess Inc., Brampton, ON, Canada, 16 p.
61. M. D. Shaw, 2008 Feedstock and Process Variables Influencing Biomass DensificationM.Sc. dissertation. Saskatoon, SK, Canada: University of Saskatchewan.
62. P. Shivanand, O. L. Sprockel, 1992 Compaction Behavior of Cellulose Polymers Powder Technology 177 184
63. Shrivastava, M., Shrivastava, P., & Khankari, K.K.1989Densification characteristics of rice husk under cold and hot compression. In Agricultural Engineering: Proceedings of the 11th International Congress on Agricultural Engineering, Dublin, 4-8 September, 2441-2443. V.A. Dodd and P. M. Grace, eds. Rotterdam: A.A. Balkema Pub4 8
64. A. Sluiter, B. Hames, R. Ruiz, C. Scarlata, J. Sluiter, D. Templeton, D. Crocker, 2008Determination of Structural Carbohydrates and Lignin in Biomass, A Technical Report: NREL/TP-510 42618National Renewable Energy Laboratory, USA, Available online at:http://www.nrel.gov/biomass/pdfs/42618.pdf
65. E. Smith, S. Probert, R. Stokes, R. Hansford, 1977 The briquetting of wheat straw Journal of Agricultural Engineering Research 105 111
66. S. Sokhansanj, S. Mani, M. Stumborg, R. Samson, J. Fenton, 2006Production and distribution of cereal straw on the canadian pariries. Canadian Biosystems Engineering, 48, 3
67. S. Sokhansanj, S. Mani, X. Bi, P. Zaini, L. G. Tabil, 2005 Binderless pelletization of biomassASAE Annual International Meeting, Tampa Convention Centre, Tampa, Florida July 17 20Paper 056061Niles Road, St. Joseph, MI 49085-9659 USA.
68. J. M. Sonnergaard, 2001 Investigation of a New Mathematical Model for Compression of Pharmaceutical Powders European Journal of Pharmaceutical Sciences 149 157
69. A. Sultana, A. Kumar, D. Harfield, 2010 Development of Agri-Pellet Production Cost and Optimum Size Bioresource Technology 5609 5621

70. L. G. Tabil, S. Sokhansanj, 1996Compression and Compaction Behavior of Alfalfa Grinds Part 1- Compression Behaviour. Journal of Powder Handling and Processing, 8(1), 17 23
71. M. Temmerman, F. Rabier, P. D. Jensen, H. Hartmann, T. Bohm, 2006 Comparative study of durability test methods for pellets and briquettes Biomass and Bioenergy 964
72. K. Theerarattananoon, F. Xu, J. Wilson, R. Ballard, L. Mc Kinney, S. Staggenborg, P. Vadlani, Z. J. Pei, D. Wang, 2011 Physical Properties of Pellets Made from Sorghum Stalk, Corn Stover, Hwaet Straw and Big Bluestem. Industrial Crops and Products 325 332
73. M. Thomas, D. J. van Zuilichem, A. F. B. van der Poel, 1997 Physical quality of pelleted animal feed. 2. Contribution of processes and its conditions Animal Feed Science and Technology 173 192
74. L. Xue, E. Mupondwa, S. Panigrahi, L. Tabil, P. Adapa, 2011Life Cycle Assessment of Biomass Densification System in Canadian Prairies. Personal Communication.
75. P. York, N. Pilpel, 1973The Tensile Strength and Compression Behavior of Lactose, Four Fatty Acids and their Mixture in relation to Tableting. Journal of Pharmacy and Pharmacology, 25, 1 11

9
CHAPTER

BEHAVIOR OF BEECH SAWDUST DURING DENSIFICATION INTO A SOLID BIOFUEL

Peter Križan*, Miloš Matúš, Ľubomír Šooš and Juraj Beniak

Institute of Manufacturing Systems, Environmental Technology and Quality Management, Faculty of Mechanical Engineering, Slovak University of Technology in Bratislava, Nám. Slobody 17, 81231 Bratislava, Slovakia

ABSTRACT

In solid biofuel manufacture technological and material variables influence the densification process and thus also the final briquette quality. The impact of these technological variables, especially compression pressure and compression temperature, and also of the material parameters (particle size and moisture content) can generally be observed during biomass densification in the quality indicators, where the abovementioned variables have a significant influence, especially on the mechanical indicators of quality (briquette density, mechanical durability, *etc.*). This paper presents the results of experimental research dealing with determining the relationship between the technological and the material variables during densification of beech sawdust. The main goal of the paper is to determine the mutual interaction between compression pressure, compression temperature and material particle size. Research findings were obtained using single-axis densification. The influence of the particle size interacting with compression pressure and compression temperature on the final briquette density was determined. The research findings obtained should prove valuable in briquette production and also in the engineering of densification machines.

Keywords: densification; briquette density; compression pressure; compression temperature; particle size

1. INTRODUCTION

The Institute of Manufacturing Systems' main interest in the environmental area is the engineering and design of densification machines and equipment. This technology is a suitable option in the treatment of biomass prior to energy recovery [1]. The Institute is attempting to develop new progressive densification machines, and trying to optimize the biomass treatment process prior to densification. The development of new machines is more effective if information about the discrete densification process is available. Requirements of biofuel production demonstrate that another important factor to take into consideration is the possibility of variable changes in the material for compression. Each different type of raw material requires a separate approach. Currently, not every densification machine can produce biofuels to the required quality according to technical standards [2]. This fact demonstrates the many problems which have to be solved. As each type of raw material has its own specific properties and chemical composition, any small change in the properties of the raw material can influence the final quality of the biofuel. Different raw material properties produce different conditions during the densification process, and this causes the final quality of the biofuel produced to be very different [3]. For each raw material it is necessary to find the optimal parameter settings for densification. This is the reason why it is so important to undertake experimental research to determine the impact of the technological and material variables on the final quality of the biofuel during densification. Briquette quality, as a final output of the densification process, is evaluated mainly by its density, and is influenced by many different variables [2]. Experience suggests that these can be divided into the following three groups [4]:

-raw material parameters (variables);

-technological parameters (variables);

-structural parameters (variables).

The general purpose of this paper is to determine the relationship between the technological and material variables, and the final quality of solid biofuels, specifically briquettes, made from beech sawdust. Initially, to evaluate if the briquettes fulfill the requirements established in the European standards, a physical characterization of each densified briquette was undertaken. Subsequently, to determine the impact of the material and technological variables, all related parameters were measured, and the densification according to the design experimental plan was carried out. Finally, in order to better understand the impact of the material and technological variables, the data was processed, and the mutual interactions between the variables investigated are described.

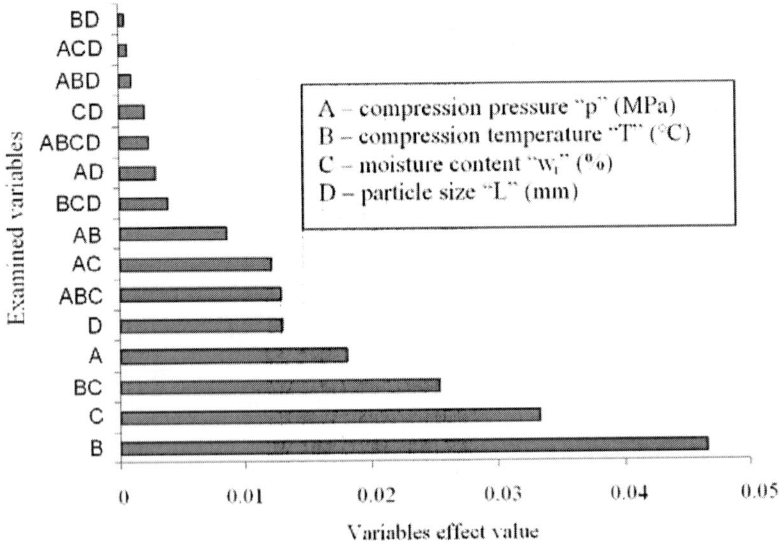

Figure 1. Paret's effect diagram for pine sawdust densification [5].

2. INVESTIGATED VARIABLES

As mentioned above, several different variables which influence the process of densification can be detected. The first group of variables relates to the type of material to be compressed; the second group relates to the densification technology; and the third group includes all the structural parameters of the machinery. Theoretical analysis of their impact significance was undertaken, and variables for the experimental research were chosen. The accuracy of these results and the appropriateness of the choice of parameters were confirmed by experimental results which were obtained from densification of pine and oak sawdust. In Figure 1 the effect of each variable during densification of pine sawdust can be seen. These variables are compression pressure "*p*", compression temperature "*T*", raw material particle size "*L*", and raw material moisture content "w_r". The goal of this research work is to determine the impact of the aforementioned variables on the quality of the final briquette.

The compression pressure is one of the most important variables affecting the resulting briquette strength [6]. With increased pressure, the briquette tends to become stronger until the yield strength of the material is reached. This also reduces the tendency of the briquette to absorb moisture when stored. It is important to find the optimum combination of force and force vector direction in order to obtain a highly compact briquette. Analysis of the pressure is interesting and complicated. To obtain a high quality briquette, it is necessary to ensure maximum pressure within the pressing chamber. This value represents the pressure needed for densification and the smooth feeding of the briquette through the chamber. Biomass under pressure undergoes two types of deformation: elastic and plastic. Elastic deformation returns to its original position once the pressure is released; however, plastic deformation permanently deforms the material. The compression temperature and pressure, and the moisture content of the material have a direct effect on the elastic and plastic deformation [5,6,7].

The compression temperature has a significant effect on the quality and the strength of the briquette [8]. Pressure and temperature go "hand in hand" in the densification process. If we increase the pressure, we can reduce the required temperature, and *vice versa*. This latter variable influences the excretion of lignin from the cellular structures [8]. Lignin is a very important substance in the densification process because of its adhesive effects in the compressed material. It connects fibres and acts as a reinforcing agent in the cellular molecules. The more lignin in the input material, and the more that can be extracted during the pressing process, the greater the uniformity and thus the quality achievable in the resulting briquette. Lignin strengthens the briquette primarily during compression, but lignin is excreted only at certain compression temperatures, which must be attained for the densification process to be optimized [9].

The optimum moisture content of the material is an essential condition for the optimal densification of the biomass and also for energy recovery

[10]. The effects of moisture content on the material must be known in order to produce quality briquettes as defined by the Standards [2]. It is known from specific research results discussed in the literature that density and strength depend on the moisture content of the compressed material, for example, in the case of compressing wood sawdust [6,7]. We can therefore say that the briquette quality depends on the moisture in the compressed material, and that a large difference is not expected in the optimal level of moisture between different materials. The optimal value, however, is not known. If the moisture content of the compressed material is very low, or very high for that matter, the particles become inconsistently arranged, and the resulting briquette becomes unstable [9,11]. Research and experiments have shown that moisture has an effect on the plasticization of lignin. The mutual interaction between the material moisture, compression pressure, and compression temperature is particularly important. If we set the process to the optimal temperature and pressure, but allow moisture content to exceed the optimal range, the resultant briquette will be unstable. The correct temperature ensures that the lignin plasticizes in the compressed material, and the optimal pressure provides densification of this material [4,8]. However, if the compressed material has a high moisture content, escaping water in the form of steam will crack the briquette. If the moisture content is very low, on the other hand, the briquette must be pressed at a higher temperature and pressure in order to obtain a high-quality product, and this becomes economically unprofitable. With reduced moisture content, the heating value must be increased. Certain moisture levels cause a higher temperature in the densification process and thus a higher briquette quality [12]. Optimal moisture level improves the compaction of the material by causing a degree of sintering which improves homogeneity of the briquette particles.

The size of the input particle also has an important effect on the densification process because larger input particles increase the energy needed for densification. Briquettes formed from large input particles also have less homogeneity and strength [7]. On the other hand, a large number of fine particles provide better material densification. The resulting briquette is uniform, of high quality, and displays higher volumetric density [11]. With increasing particle size, the bond strength between particles decreases, causing them to crumble [9]. As a result, the quality of the briquette decreases and the compression pressure necessary to counteract this drop in quality increases. During compression, especially without additives, the surfaces of the particles must have contact over the greatest possible area [7]. The size of this contact area increases as the particle size becomes finer and higher compression pressures need to be applied. In the material, particles exist which deform and load mainly the contact regions, due to external forces. In terms of densification, it is very important that bonding forces are created between the particles. The strength of the resultant bonds increases with decreasing particle size. However, the optimum particle size changes in relation to the densification technology employed, and the raw material used. Therefore, it is important to investigate the effects of particle size on the resulting density of the briquette. Thus particle size was one of the variables that this research attempted to identify and verify in the experiments.

3. EXPERIMENTAL RESEARCH CONDITIONS

European beech (*Fagus sylvatica*) sawdust with a moisture content of 8%, without bark, and originating from Western Slovakia was obtained from a wood processing company. Initially, the particle size distribution was analyzed using an AS 200 Vibrating Sieve Shaker (Retsch, Haan, Germany). The raw material particle size distribution was calculated by weight, see Figure 2. For the experimental research particle sizes 0.5, 1.0, 2.0, 4.0 mm and an unsorted mix were chosen, (see Figure 3).

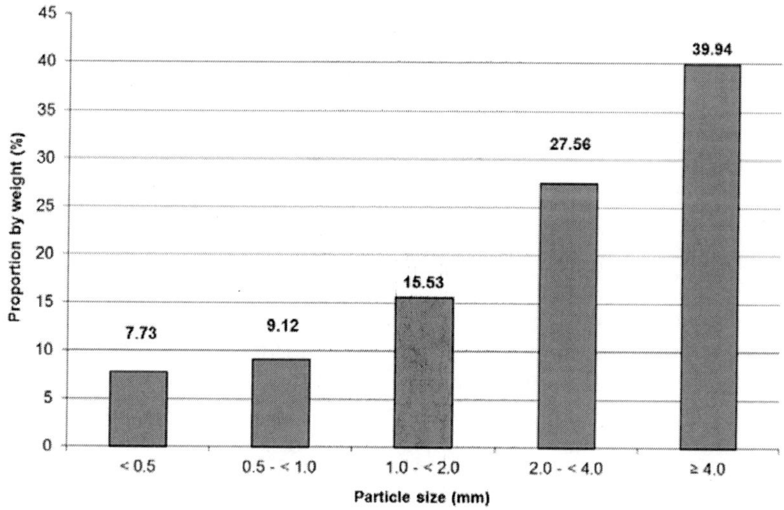

FIGURE 2. Raw material particle size distribution.

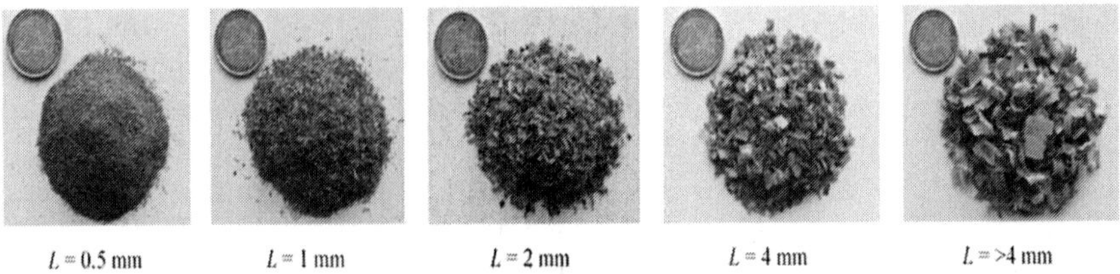

FIGURE 3. Beech sawdust particle sizes.

The moisture content of the beech sawdust was measured prior to undertaking the experimental research using a MRS 120-3 balance (Kern, Balingen, Germany). This measurement was based on heating the material (gravimetric method of moisture content measuring) at 105 ± 2 °C until a constant weight was achieved [13]. Due to the complicated moisture content change process and time limitations, only one moisture content level was used in this phase of the experiments (w_r = 8%), so the impact of moisture content on briquette quality was not investigated or calculated in our experiments.

Briquettes were produced using a vertical hydraulic press which was modified with the use of an experimental pressing stand (see Figure 4). The experimental pressing stand consists of a base frame, a cylindrical pressing chamber with a 20 mm diameter die, a heating device with a temperature sensor for temperature control, and the backpressure plug. When the temperature in the chamber was reached, the raw material was fed into the chamber and compression with a hydraulic piston commenced. The hydraulic press made it possible to set the compression pressure in the range 31 MPa to 318 MPa. Adhering to the experimental plan for each experiment setting, five briquettes were produced. This number was chosen to satisfy standard mathematical and statistical methods for experiment evaluation [4]. The described hydraulic press with stand allowed experimental settings over a wide range of compression pressures and temperatures. Densification of the beech sawdust was increased gradually in order to achieve the optimum results.

FIGURE 4. Experimental pressing stand.

Briquette density was calculated using the ratio between briquette weight and its volume. The weight, length and diameter of each briquette were measured with a digital caliper and an electronic balance. The volume of the briquettes was calculated as the volume of a cylinder with dimensions (length and diameter), according to EN 16127 [14]. Average density was calculated for each sample briquette. All the results in this phase, showing the interrelation between the different variables, were depicted graphically, as seen in the figures below. The results of this experimental research are applicable to the choice of dimensions of densification machines, and gave a comprehensive overview of the behavior of different variables during the densification process.

4. RESULTS AND DISCUSSION

The impact of compression pressure on final briquette density can be seen in Figure 5. As the compression pressure increases so too does the briquette density. This interdependence also highlights the importance of briquette stabilization time.

Briquette stabilization time is the time interval during which dilatation occurs; it is also the time interval during which the briquette stabilizes. Briquette stabilization time takes approximately 24 h depending on the type of compressed material and the densification technology used, but it can also take longer. The DIN 52182 Standard [2] describes the conditions and process for detecting briquette density after stabilization time. After compression, the briquettes had to be placed under stable climatic conditions.

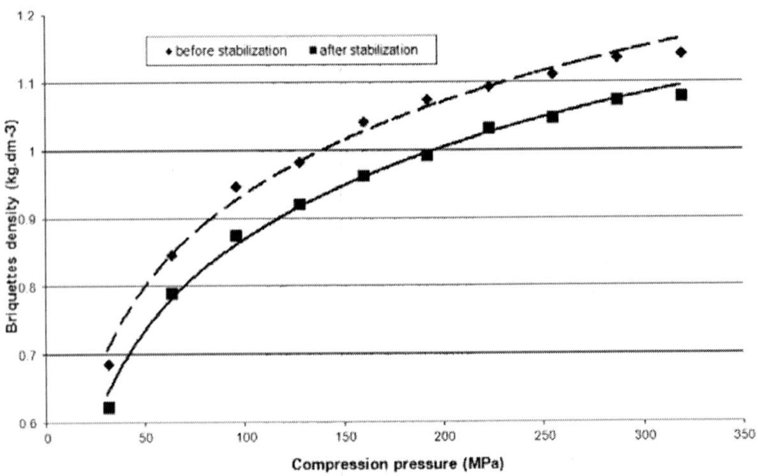

FIGURE 5. Briquette density of beech sawdust, related to compression pressure at 25 °C (w_r = 8%; L = 2 mm) [4].

During this stabilization time the briquette dimensions (length and diameter) were frequently measured and its weight and density were evaluated. If the briquette weight had changed by no more than a maximum of 0.1% during the last 24 h, the briquette was considered stable [2]. Briquette dilatation is a process during which the briquette's dimensions (diameter, length) and weight undergo a change. These changes result from the internal variables of the briquette material, and also from the external variables of the densification technology. Dilatation has a direct influence on the briquette density because the density was calculated from the aforementioned briquette dimensions. The dilatation effect is produced when the compression pressure is removed, and, as a result of the relationship of the interactions between the compression temperature, material moisture, and input particle size.

Two of these interdependencies can be seen in Figure 5, one for densities evaluated immediately after releasing the briquettes from the compression chamber, *i.e.*, before stabilization, and the second for densities evaluated after the stabilization time. The briquette density changed significantly, and dilatation negatively influenced the briquette density in this case.

This effect can significantly distort the results of the experiment. Briquette density is important from the point of view of material particle cohesion. During densification, lignin, which acts as a natural glue, is released from the biomass. If the briquette is not created under optimally adjusted conditions, dilatation of the briquette can also significantly influence the absorption of moisture from its surroundings. In practice, the stabilization time can be reduced by cooling. In Figure 6, Figure 7 and Figure 8, it can be seen that each briquette produced from the same settings, and also those from other settings were subject to dilatation during the stabilization time. The size of dilatation depends on the variables investigated here. Increasing the compression pressure positively affects the degree of dilatation; *i.e.*, dilatation is reduced. We can see that the differences between individual briquette density values before and after stabilization are reduced when the compression pressure is increased.

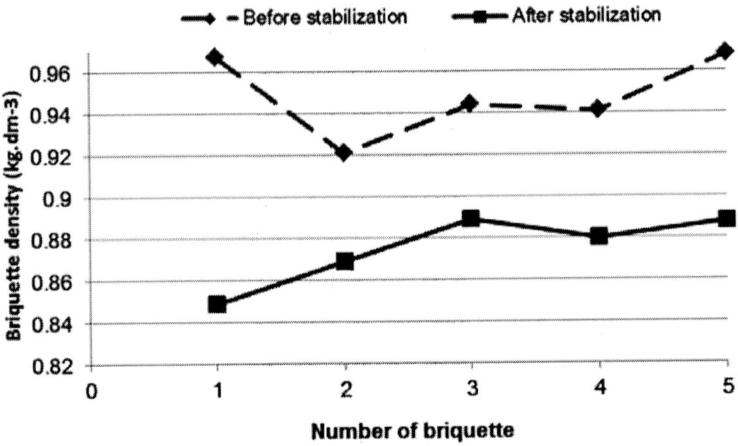

FIGURE 6. Briquette density values at 95 MPa compression pressure (w_r = 8%;L = 2 mm).

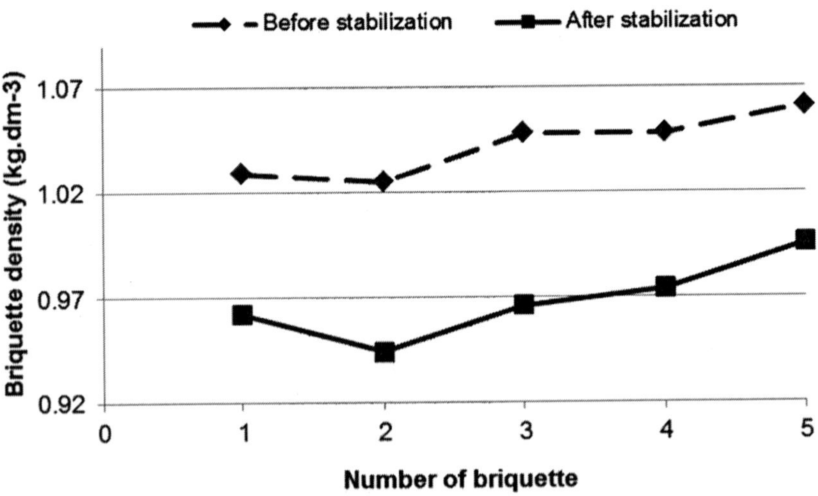

FIGURE 7. Briquette density values at 159 MPa compression pressure (w_r = 8%; L = 2 mm).

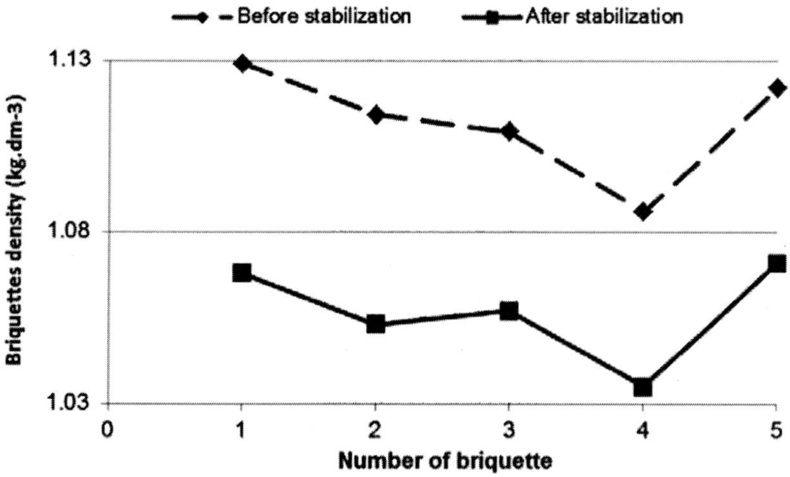

FIGURE 8. Briquette density values at 254 MPa compression pressure (w_r = 8%; L = 2 mm).

In Figure 9, the interdependence between differing densities, both before and after stabilization, and compression pressure, can be observed graphically. This dependence clearly evidences the impact of compression pressure on briquette dilatation and means that with increasing compresson pressure, briquette dilatation is reduced.

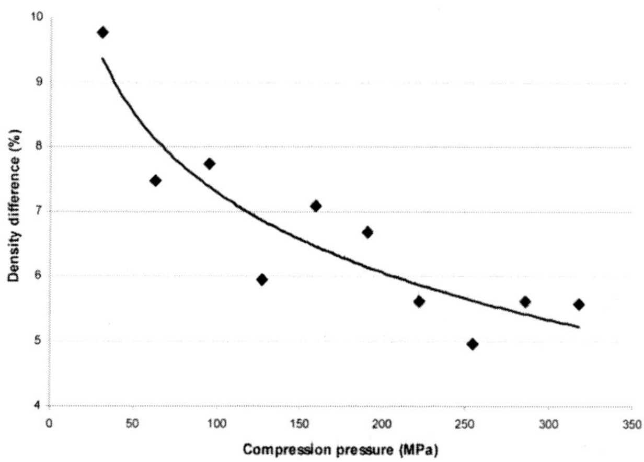

FIGURE 9. Density difference, dependent on pressing pressure (w_r = 8%; L = 2 mm).

A significant part of the experimental research was also studying the effect of raw material particle size on briquette density. The influence of particle size on briquette density (see Figure 10) has shown that reducing the particle size positively influences the increase in briquette density.

The figure shows that with increasing particle size, briquette density changes. Also it is important to note that for these results, the measurements were taken while compacting without any increase in temperature during the process, and at three levels of compression pressure.

In the case of particle size, the beech sawdust compressibility factor is an important variable. The compressibility factor characterizes one of the basic properties of sawdust—compression. Sawdust compression is defined as the tendency of bulk sawdust to change volume under the effect of an external force. This means that the magnitude of this compressibility or compression factor is affected by the pressure acting upon it. The results of the experiments show that increasing the compression pressure for beech sawdust of equal particle size increases the briquette density (see Figure 17 below). Also it was proven that by increasing the particle size the briquette density decreases. These results imply that important variables affecting the compression factor are the pressure applied and the particle size.

These two variables also influence the friction conditions in the pressing chamber and thus also the force needed to eject the briquette from the pressing chamber. The process of densification of biomass is caused by the piston moving from one end of the chamber, and by the frictional resistance of the briquettes through the chamber. The increased friction, depending on the length of the chamber, causes a higher compression pressure on the raw material. The determination of the force needed to eject the briquette from the chamber depending on the material and technological variables chosen was a secondary objective of this research.

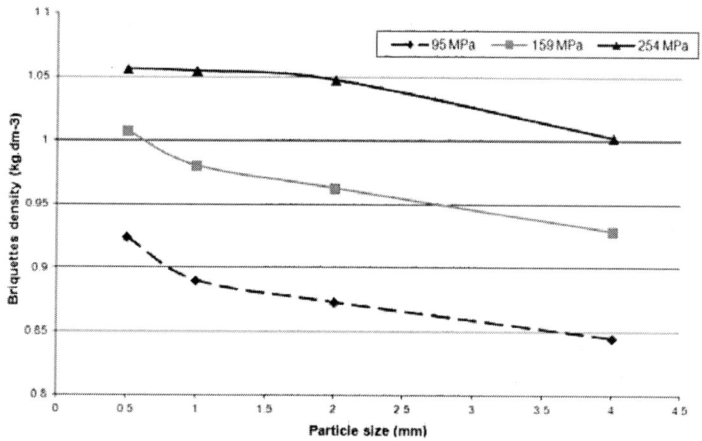

FIGURE 10. Dependence of briquette density on particle size (w_r = 8%; T = 25 °C).

The interdependence of force required for ejection and particle size can be seen on Figure 11. During the experiments, briquettes of differing lengths were produced. Therefore the force required to eject the briquette was recalculated to provide a "per 1 mm length of briquette" figure. Increasing particle size while increasing compression pressure also increases the force needed for briquette eject.

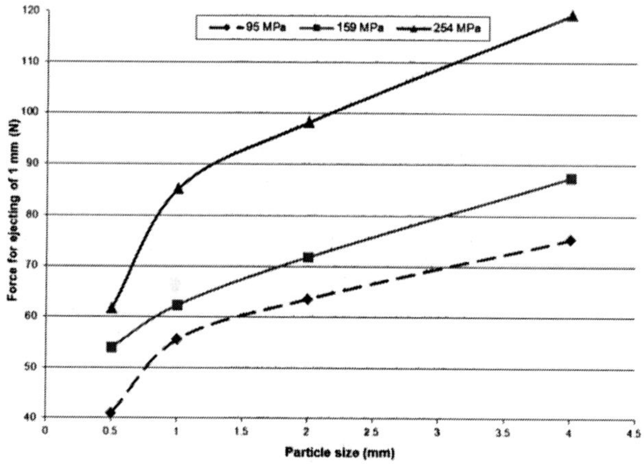

FIGURE 11. Interdependence of force needed to eject the briquette and particle size (w_r = 8%; T = 25 °C).

These findings should be combined with the impact on briquette density. The results are presented in Figure 12, Figure 13 and Figure 14. Here the cumulative interdependencies can be seen in graphic form. These show that the degree of force needed for ejecting and briquette density depend on particle size. For economic efficiency, it is very important to know how much the raw material needs to be pulverized in order to obtain the optimal particle size.

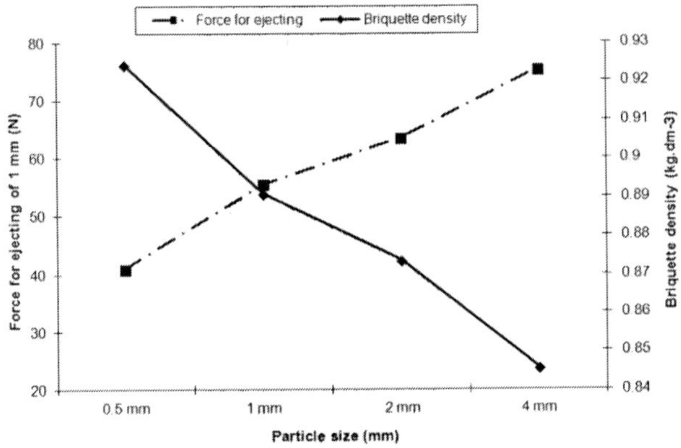

FIGURE 12. Cumulative interdependence—force required for ejecting, briquette density, and particle size (w_r = 8%; T = 25 °C) at 95 MPa.

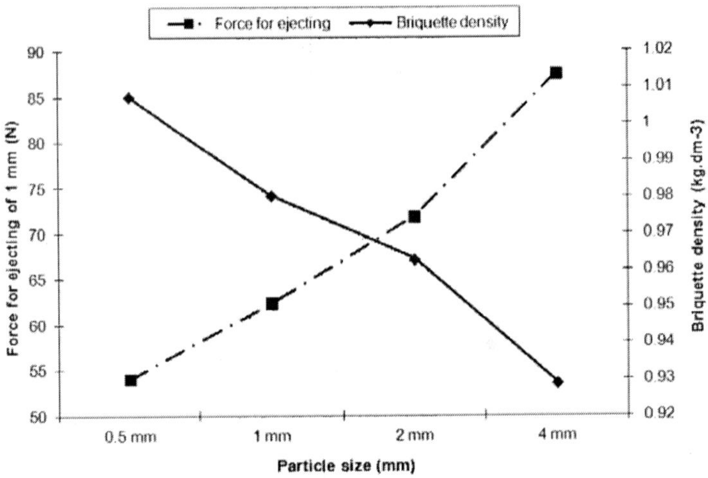

FIGURE 13. Cumulative interdependence—force required for ejecting, briquette density, and particle size (w_r = 8%; T = 25 °C) at 159 MPa.

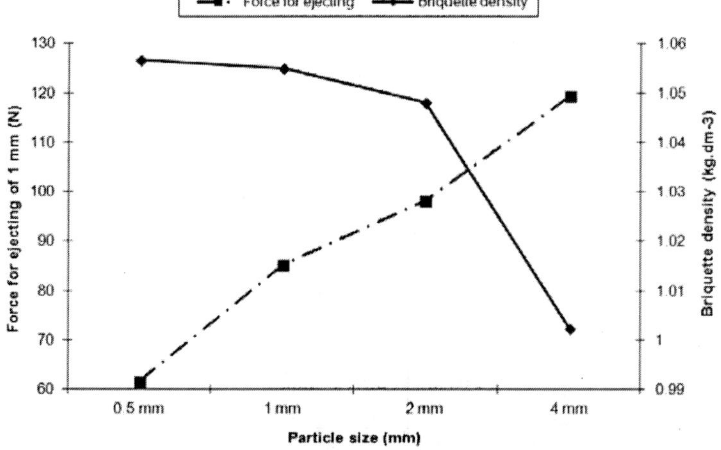

FIGURE 14. Cumulative interdependence—force for ejecting briquette, briquette density, and particle size (w_r = 8%; T = 25 °C) at 254 MPa.

The influence of compression pressure and particle size can be very clearly seen also in Figure 15 and Figure 16. Briquettes with 20 mm diameter were produced and measured during the experimental research. Table 1presents views which were obtained during the investigation of beech briquette surfaces using a digital microscope. All views were obtained with the same zoom of 40×. Different particle deposition can be seen in the briquettes after densification, according to different particle size. A smoother briquette surface structure positively influences both the briquette density and also the frictional resistance of the briquette in the pressing chamber. When no compression temperature was applied, it can be seen that only bridges of a solid nature exist as a mechanism for internal bonding. When compression temperature is applied, this causes lignin plasticization and the creation of liquid bridges between the material particles.

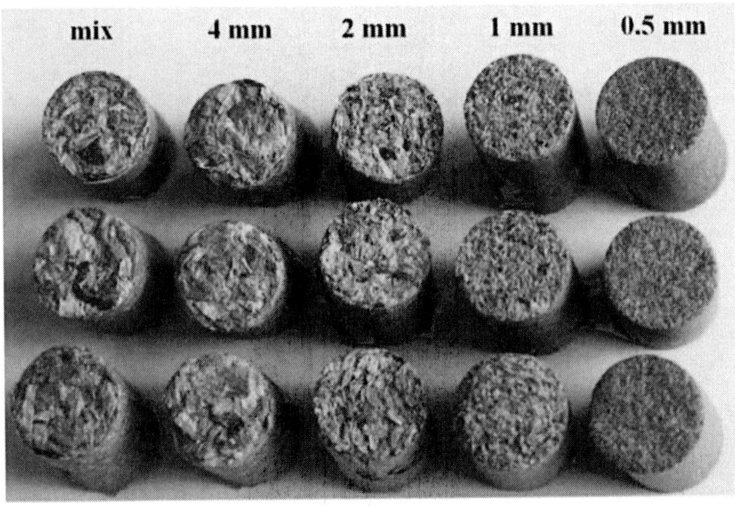

FIGURE 15. Comparison of beech briquettes from various particle sizes (p = 159 MPa; w_r = 8%; T = 25 °C).

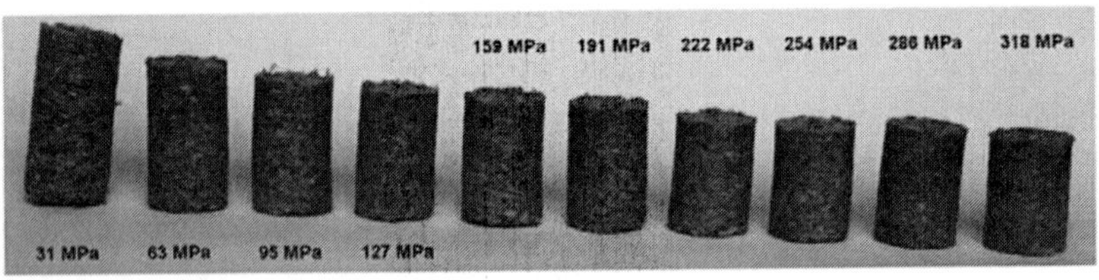

FIGURE 16. Briquettes from beech sawdust with different pressures applied (w_r = 8%; T = 25 °C).

The interdependency of briquette density and compressive pressure during densification of different particle sizes can be seen in Figure 17. It was proven that by reducing the particle size and increasing the compression pressure, briquette density also increases. This result demonstrating the interaction between particle size and compression pressure is extremely significant from the point of view of friction. The interactions investigated influence the binding forces produced between the material particles. How the friction changes with the particle size and the changing compression pressure can be clearly seen in Figure 18. Reducing the particle size also reduces the force needed to eject the briquette. This force represents the friction between the pressed material and the chamber wall.

TABLE 1. Zoom views of the investigated briquette surfaces using a digital microscope (40×).

Particle Size (mm)	View Perpendicular to the Briquette Axis	View Along the Briquette Axis
0.5	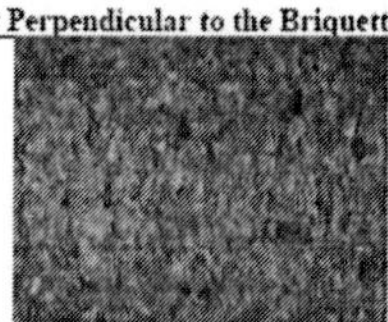	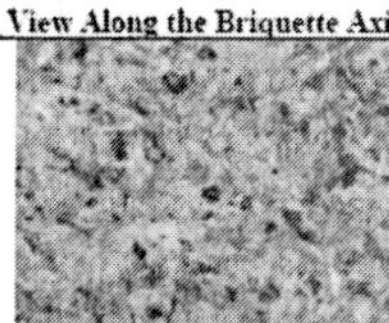
1.0		
2.0		
4.0	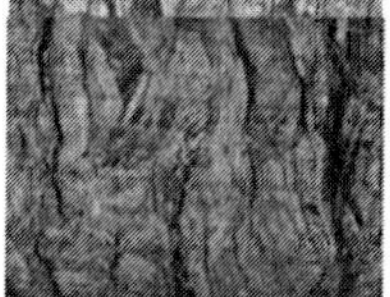	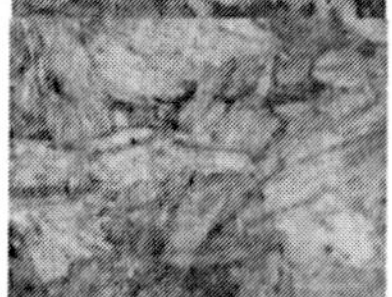
mix		

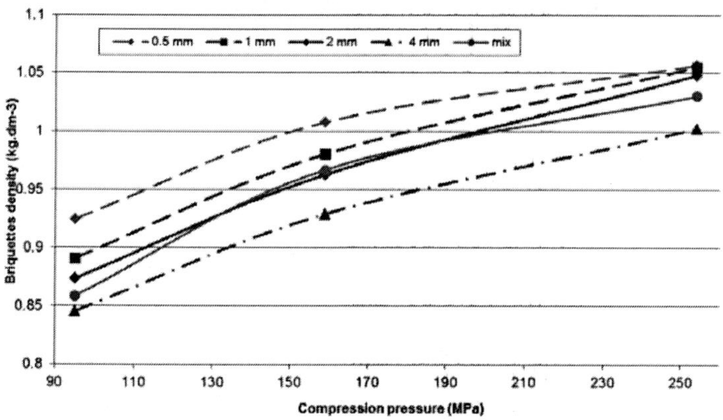

FIGURE 17. Relation between briquette density and compression pressure for different particle sizes (w_r = 8%; T = 25 °C).

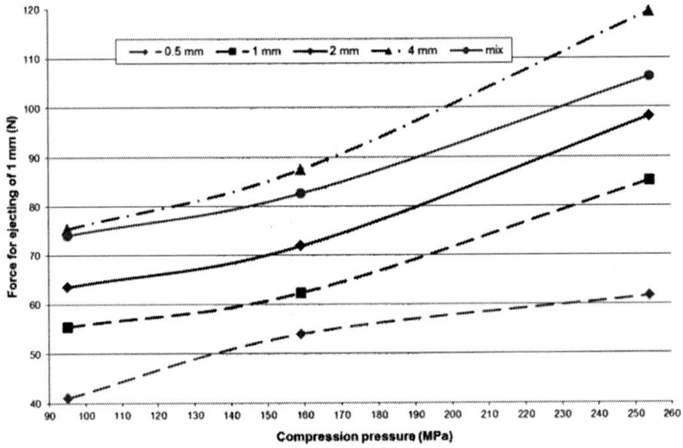

FIGURE 18. Relation between force needed for ejection and compression pressure for different particle sizes (w_r = 8%; T = 25 °C).

Briquette density and thus also the frictional resistance of the briquettes has a significant influence on compression temperature. The previous results were obtained at a compression temperature of 25 °C. The results presented in Figure 19 were obtained as a result of densification at three other different pressing temperature levels (55, 85 and 115 °C). Figure 19 shows the relationship between briquette density the compression pressure at four different levels of compression temperatures. The positive impact of compression temperature on the final briquette quality was proven. When the compression temperature and the compression pressure are increased, briquette density increases. The compression temperature has a positive impact on the lignin component of the compressed material [9]. The compression temperature influences lignin plasticization [8,9]. With increased compression temperature during densification, the lignin becomes more and more liquid, plastic. Temperature significantly affects the bonding mechanism forming between the solid particles of the compressed material [8,9]. A more effective bonding of the cellular structure of the compressed material and the required plasticization of lignin is possible only under induced compression temperatures.

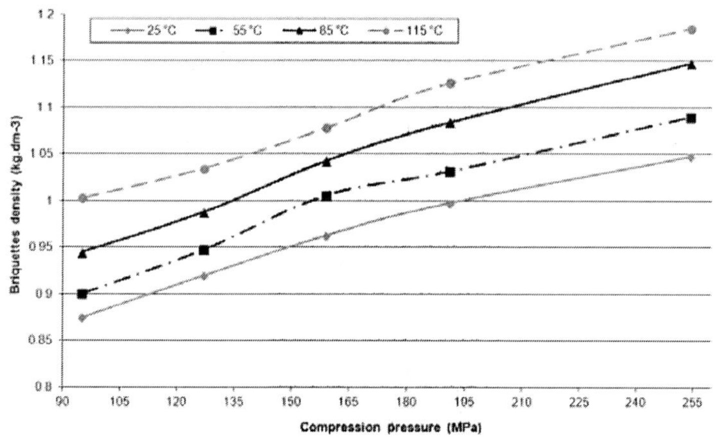

FIGURE 19. Relation between briquette density and compression pressure at different compression temperatures (w_r = 8%; L = 2 mm) [8].

5. CONCLUSIONS

Research into beech sawdust densification as a wood waste was investigated in this research. The main conclusions that can be drawn from this study are as follows:

- All of the investigated variables have a significant effect on final briquette density. Increased compression pressure and compression temperature, and reduced particle size, increase the density of beech briquettes.
- The density value laid down in the technical standard (1.14 kg·dm^{-3}) [2] can be achieved only under induced compression temperatures.
- A group of parameters (particle size, compression pressure, compression temperature) significantly affect the force needed to eject briquettes from the compression chamber and because of the friction between the briquette and the chamber wall.

ACKNOWLEDGMENTS

The research presented in this paper is an outcome of project No. APVV-0857-12 “Tools durability research into progressive compacting machine design and development of adaptive control for the compaction process” funded by the Slovak Research and Development Agency.

AUTHOR CONTRIBUTIONS

Individual contributions of all authors are as follows: Peter Križan and Ľubomír Šooš analyzed the initial state and designed the experiments; Miloš Matúš and Juraj Beniak prepared all laboratory equipments and raw material for experiments; Miloš Matúš and Peter Križan performed the experiments; Juraj Beniak and Peter Križan analyzed the obtained data; Peter Križan and Ľubomír Šooš wrote the paper.

REFERENCES

1. Tchapda, A.H.; Pisupati, S.V. A review of thermal co-conversion of coal and biomass/waste. *Energies* **2014**,*7*, 1098–1148.
2. *Bestimmung der Rohdichte*; DIN 52182; Deutsches Institut für Normung: Berlin, Germany, 1976. (In German)
3. Qian, K.; Kumar, A.; Patil, K.; Bellmer, D.; Wang, D.; Yuan, W.; Huhnke, R.L. Effects of biomass feedstocks and gasification conditions on the physiochemical properties of char. *Energies* **2013**, *6*, 3972–3986.
4. Križan, P.; Šooš, L.; Vukelič, Đ. A study of impact technological parametres on the briquetting process. *Sci. J. Facta Univ.* **2009**, *6*, 39–47.
5. Križan, P. *The Densification Process of Wood Waste*, 1st ed.; De Gruyter Open: Berlin, Germany, 2015; p. 184.
6. Kaliyan, N.; Morey Vance, R. Factors affecting strength and durability of densified biomass products.*Biomass Bioenergy* **2009**, *33*, 337–359.
7. Mani, S.; Tabil, L.G.; Sokhansanj, S. Effects of compressive force, particle size and moisture content on mechanical properties of biomass pellets from grasses. *Biomass Bioenergy* **2006**, *30*, 648–654.
8. Križan, P.; Svátek, M.; Matúš, M.; Beniak, J. Impact of pressing temperature on the pressing conditions in briquetting machine pressing chamber. *J. Prod. Eng.* **2014**, *17*, 79–82.
9. Nielsen, N.P.K.; Gardner, D.J.; Poulsen, T. Importance of temperature, moisture content, and species for the conversion process of wood residues into fuel pellets. *Wood Fiber Sci.* **2009**, *41*, 414–415.

10. Arce, M.E.; Saavedra, A.; Míguez, J.L.; Granada, E.; Cacabelos, A. biomass fuel and combustion conditions selection in a fixed bed combustor. *Energies* **2013**, *6*, 5973–5989.
11. Thoreson, C.P.; Webster, K.E.; Darr, M.J.; Kapler, E.J. investigation of process variables in the densification of corn stover briquettes. *Energies* **2014**, *7*, 4019–4032.
12. Matúš, M.; Križan, P. Influence of structural parameters in compacting process on quality of biomass pressing. *J. Appl. Math.* **2010**, *3*, 87–96.
13. European Committee. *Solid Biofuels—Determination of Moisture Content-oven Dry Method Part 3: Moisture in General Analysis Sample*; EN 14774-3:2009; European Committee for Standardization: Brussels, Belgium, 2009.
14. European Committee. *Solid Biofuels—Determination of Length and Diameter of Pellets*; EN 16127:2012; European Committee for Standardization: Brussels, Belgium, 2009.

10

CHAPTER

BIOFUEL PRODUCTION IN IRELAND—AN APPROACH TO 2020 TARGETS WITH A FOCUS ON ALGAL BIOMASS

Fionnuala Murphy[1], Ger Devlin[1,*], Rory Deverell[2] and Kevin McDonnell[1]

[1] *School of Biosystems Engineering, University College Dublin, Belfield, Dublin 4, Ireland*

[2] *R&H Hall, La Touche House, Custom House Dock, IFSC, Dublin 1, Ireland*

ABSTRACT

Under the Biofuels Obligation Scheme in Ireland, the biofuels penetration rate target for 2013 was set at 6% by volume from a previous 4% from 2010. In 2012 the fuel blend reached 3%, with approximately 70 million L of biodiesel and 56 million L of ethanol blended with diesel and gasoline, respectively. Up to and including April 2013, the current blend rate in Ireland for biodiesel was 2.3% and for bioethanol was 3.7% which equates to approximately 37.5 million L of biofuel for the first four months of 2013. The target of 10% by 2020 remains, which equates to approximately 420 million L yr−1. Achieving the biofuels target would require 345 ktoe by 2020 (14,400 TJ). Utilizing the indigenous biofuels in Ireland such as tallow, used cooking oil and oil seed rape leaves a shortfall of approximately

12,000 TJ or 350 million L (achieving only 17% of the 10% target) that must be either be imported or met by other renewables. Other solutions seem to suggest that microalgae (for biodiesel) and macroalgae (for bioethanol) could meet this shortfall for indigenous Irish production. This paper aims to review the characteristics of algae for biofuel production based on oil yields, cultivation, harvesting, processing and finally in terms of the European Union (EU) biofuels sustainability criteria, where, up to 2017, a 35% greenhouse gas (GHG) emissions reduction is required compared to fossil fuels. From 2017 onwards, a 50% GHG reduction is required for existing installations and from 2018, a 60% reduction for new installations is required.

Keywords: biodiesel; bioethanol; microalgae; macroalgae; indigenous production; Ireland

1. INTRODUCTION

The cultivation of algal biomass for the production of third generation biofuels has received increasing attention in recent years as they can be produced in the marine environment and on non-arable lands, thus bypassing the food *versus* fuel debate. There are two types of algae: seaweeds (macroalgae) and phytoplanktons (microalgae).

Algae have been receiving attention as a potential biofuel feedstock as they have a number of advantages over traditional biofuel crops:

- Production yields of algae per unit area are significantly higher than those for terrestrial biomass [1,2].
- The chemical composition of algae makes it suitable for conversion into biofuels [3,4,5,6]. In general, microalgae are potential sources of bio-oils whilst macroalgae are potential sources of carbohydrates for fermentation or thermo-chemical based conversions [7].
- Microalgae can be cultivated in brackish water on non-arable land, and therefore may not incur land use change, minimizing associated environmental impacts [6]. Utilizing marine biomass, which can be grown in a variety of marine environments including fresh water and salt water, avoids the problem of land use change from arable to bioenergy crops [1,8]. Utilizing the marine environment ensures a large cultivation area, limiting competition with other land uses and resources [3].

- Microalgae production can utilize the carbon dioxide component of flue gas, reducing the carbon emissions from power plants [2]. Marine macroalgae have a high rate of carbon dioxide fixation from the atmosphere and water at 8–10 t ha^{-1} yr^{-1}, comparable to temperate woodlands [3,4]. As such, they have a high potential for carbon dioxide remediation.

The main steps in algae production; cultivation, harvesting and conversion to biofuel, are discussed separately for microalgae and macroalgae since there are intrinsic differences between them in all aspects of the process.

2. EUROPEAN UNION (EU) SUSTAINABILITY CRITERIA

The increased use of biomass for biofuel production has led to concerns regarding the sustainability of this practice. Concerns surround the methods of cultivating and producing biofuels, particularly in regard to actual greenhouse gas (GHG) emissions reductions in comparison with fossil fuels, and in relation to land use change due to increased demand for arable land for biomass production. In order to ensure the sustainability of biofuel used to achieve the targets in the EU, the European Commission proposed a set of sustainability criteria in the Directive 2009/28/EC on the promotion of the use of energy from renewable sources. The sustainability criteria consist of the following main points:

- The directive lays out certain GHG emissions reductions to be achieved from the use of biofuels. In the case of biofuels and produced by installations that were in operation on 23 January 2008, GHG emissions savings must be at least 35% from 2013. This figure rises to 50% in 2017, and further to 60% in 2018 for biofuels produced in installations in which production started on or after January 2017.
- The raw materials sourced for biofuel production, from within the EU or from third countries, should not be obtained from land with high biodiversity value, land with a high carbon stock, or land that was peatland in 2008 [9].

These criteria, while undoubtedly positive for the sustainable production of biofuels, may restrict growth of the biofuel production industry in Ireland as biofuels must meet these minimum criteria.

3. MICROALGAE

Microalgae are microscopic algae that typically range from unicells to colonies and filaments of up to a few hundred cells. Microalgae include prokaryotes (cyanobacteria and blue-green algae) and eukaryotes (green algae, diatoms, red algae and others) [10]. There are thousands of microalgae species, only some of which have been studied for biofuel production.

Microalgae are particularly suited to biofuel production due to their high photosynthetic growth rates, high lipid content, low land usage and high carbon dioxide absorption [11]. Algae absorb freely available sunlight and can utilize waste streams to provide essential nutrients for cultivation. Algae can convert waste CO_2 from power plant exhaust gas to organic biomass which can then be converted to energy [12]. Moreover, municipal wastewater streams can be harvest to provide additional nutrients [13].

Microalgae are already reported to produce 15–300 times more oil for biodiesel production than traditional crops on an area basis (Table 1). Furthermore compared with conventional crop plants which are usually harvested once or twice a year, microalgae have a very short harvesting cycle (1–10 days depending on the process), allowing multiple or continuous harvests with significantly increased yields [2].

TABLE 1. Comparison of crop-dependent biodiesel production efficiencies from plant oils [2]. TAG: Triacylglycerols.

Plant source	Biodiesel ($L\ ha^{-1}\ yr^{-1}$)	Area required to produce global oil demand (hectares × 10^6)	Area required as percent global land mass	Area as percent global arable land
Cotton	325	15,002	100.7	756.9
Soybean	446	10,932	73.4	551.6
Mustard seed	572	8,524	57.2	430.1
Sunflower	952	5,121	34.4	258.4
Rapeseed/canola	1,190	4,097	27.5	206.7
Jatropha	1,892	2,577	17.3	130
Oil palm	5,950	819	5.5	41.3
Microalgae ($10\ g\ m^{-2}\ day^{-1}$ at 30% TAG)	12,000	406	2.7	20.5
Microalgae ($50\ g\ m^{-2}\ day^{-1}$ at 50% TAG)	98,500	49	0.3	2.5

3.1. Cultivation and Harvest

There are two main methods for cultivating microalgae; raceway pond systems and photobioreactors. A typical raceway pond comprises a closed loop oval channel, approximately 0.25–0.4 m deep, open to the air, and mixed with a paddle wheel to circulate the water and prevent sedimentation. Ponds are kept shallow as optical absorption and self-shading by the algal cells limits light penetration through the algal broth [14]. In photobioreactors, the culture medium is enclosed in a transparent array of tubes or plates and the micro-algal broth is circulated from a central reservoir. Productivity is higher in the controlled, contained environment of a photobioreactor, but capital and operating expenses are also substantially higher than for open systems [6]. Auxiliary energy demand may also be higher [14].

Algae typically have a high water content and downstream harvesting and processing requires its removal [15]. In existing algal aquaculture the most common harvesting processes are flocculation, microscreening and centrifugation. Most importantly, cost-effective and energy-efficient harvesting methods are required to make the entire biofuels production process economical [2].

Cell disruption is an important step in recovering intracellular products from micro-algae, and so properties of the cell wall play an important role in the extraction process. Some of the commonly used methods for cell disruption include mechanical disruption, such as bead-beating, ultrasound and steam extraction [16] and non-mechanical disruption, including application of organic solvents and addition of inorganic acids and alkali for pre-treatment processing [17]. For extraction of oils and other microalgal products, chemical solvents can be chosen in one or two-step extraction approaches [2]. The residue remaining after oil extraction contains starch and proteins and can be further processed to produce ethanol, animal feed, or used as a feedstock for anaerobic fermentation [18].

3.2. Processing

The optimal usage is highly dependent on the microalgal composition, for example species with high lipid content will preferably be used for biodiesel production, while algae high in carbohydrates are more suitable for bioethanol. Combinations are also possible, in which for instance the lipid fraction is used for biodiesel, while the remaining biomass is fermented for biogas production. Biorefinery-type processes may be the most economical method of producing algal biofuels as several commercial products can be obtained from the algal biomass. Lipids can be extracted for biodiesel production, while other products can be fermented to produce ethanol and biogas. It is also possible to produce protein-rich feed for both animal and human consumption [19]. Bulk markets for the co-products are potentially available [20]. The common biochemical and thermochemical processes are outlined in Figure 1.

3.3. Biochemical Processing

3.3.1. Transesterification

The transesterification process involves reaction of an alcohol with triglycerides, in the presence of a catalyst, forming fatty acid alkyl esters. The transesterification process can be acid or base catalyzed, and involves enzymatic conversion. In acid-catalyzed reactions, HCl, H_2SO_4 or H_3PO_4 is used for transesterification, while in base catalysis strong bases like KOH or NaOH are commonly used. Base catalysis has several advantages over the acid-catalyzed reaction as it is conducted at low temperature and pressure, and it has also a high conversion rate. Despite these advantages, the process is energy intensive, and there are problems associated with removal and treatment of alkaline catalyst from the final product. These problems could be solved by the use of biocatalysts, such as lipases, but large-scale demonstration has not been reported [17].

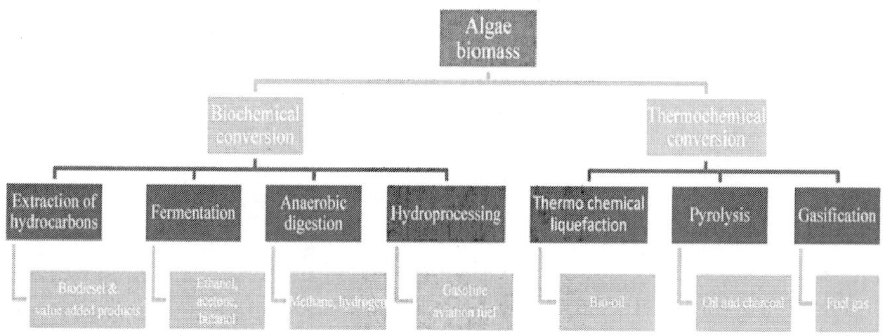

FIGURE 1. Potential algal biofuel conversion processes, adapted from [21].

3.3.2. Fermentation

Microalgae are also a suitable feedstock for bioethanol production. Besides their high lipid content, some microalgae also contain carbohydrates (generally not cellulose) that can be utilized as a carbon source or substrate for fermentation. Certain algae accumulate starch in stress conditions, and this can be exploited for ethanol production by fermentation. Ethanol can be produced either from algal biomass or from algal cake. The processes for producing bioethanol from microalgae are very similar to establishing first-generation technologies that use corn- and sugarcane-derived feed stocks [2]. Despite these possibilities, research on the fermentation of microalgae biomass to bioethanol is still limited in the literature [19].

3.3.3. Anaerobic Digestion

Anaerobic digestion is a potential option for conversion of algal residue to energy. The anaerobic digestion of algae can achieve methane yields of about 250 $m^3\,t^{-1}$ of algae. The three main bottlenecks to digest microalgae are; low biodegradability of microalgae depending on both the biochemical composition and the nature of the cell wall, high cellular protein content results in ammonia release which can lead to potential toxicity, and the presence of sodium for marine species can also affect the digester performance [22].

3.3.4. Hydroprocessing

In recent years, hydroprocessing technology has been used to convert lipid feedstocks into distillate fuels [20]. The two main products consist of a liquid distillate fuel with similar properties to petroleum diesel, and propane in the gas phase stream. With the various hydroprocessing technologies utilised by refineries to catalytically remove impurities or reduce molecular weight, algal oils can be processed into a kerosene-like fuel very similar to petroleum-derived jet fuels [23]. This fuel is being considered as a jet fuel for several reasons; the cold weather properties are superior, the propane byproduct is preferable over glycerol byproduct, the heating content is greater, and the cetane number is greater [24]. The main advantage of hydroprocessing is that it requires infrastructure which is widely available in all refinery units [25].

3.4. Thermochemical Processing

3.4.1. Hydrothermal Liquefaction

An alternate approach that requires no feedstock drying and no use of organic solvents is hydrothermal liquefaction. This approach converts the microalgae, with high moisture content, in water at temperatures of

250–350 °C and high pressure (5–20 MPa) [26], and has been investigated for various strains (*Botryococcus braunii,Spirulina platensis, Chlorella vulgaris, Nannochloropsis sp.* or *Desmodesmus sp.*, among others) [27,28,29]. This process is particularly suited for high moisture feedstocks, overcoming one of the key barriers for other processing techniques [30]. Under these conditions, the biomacromolecules in the microalgae break down to form a liquid energy carrier called "bio-oil" or "biocrude", next to gaseous, aqueous and solid by-products [27,31]. The biocrude produced, with an energy value close to that of fossil petroleum [30] is not directly suited as transportation fuel, but it is expected to be a suitable renewable feedstock for co-refining in existing fossil-based refineries [31]. Maximum biocrude yields in the vicinity of 50%–60% have been reported, and the use of several homogeneous and heterogeneous catalysts has explored [28,32], though in a very early stage of development.

3.4.2. Gasification

Gasification is a thermochemical process that, in the near absence of oxygen, converts organic material into a combustible gas called synthesis gas (syngas) [22]. Gasification of microalgae can result in production of clean $H_2$with yields ranging from 5% to 56% and CO with yields ranging from 9% to 52% [33]. Methane can be considered to be a co-product and is only produced in small amounts of approximately 2%–25% [33]. However, the production of clean methane-rich gas can be achieved in the catalyzed supercritical water gasification process where approximately 60%–70% of the heating value from the microalgal biomass can be recovered as methane. The hydrocarbon products of gasification can be further processed to produce methanol. At 1000 °C, the methanol production is approximately 64% (*w/w*) based on the biomass weight [34]. However, methanol production is an energy intensive process, due to the use of the centrifuge process during harvesting, with the ratio of energy produced to energy required of 1.1 [35]. The biomass gasification also produces unwanted products in small quantities such as water, ash and tar, which cause various problems with the main product yield. The tar produced can range from 0.1% to 20% depending on the gasifier agent and type of reactor, either an updraft or a downdraft reactor [33].

3.4.3. Pyrolysis

Pyrolysis is accomplished at moderate to high temperatures (400–600 °C) and atmospheric pressure and requires drying of the feedstock. Slow pyrolysis is the heating of biomass at slow heating rates (5–80 °C min^{-1}) and longer residence times (5–30 min), compared to high heating rate (1000 °C min^{-1}) and short residence times (10–20 s) in case of fast pyrolysis [30]. Bio-oil yields of 17.5% and 23.7% were reported from the fast pyrolysis of*C. protothecoides* [36] and *M. aeruginosa* [37]. The bio-oil yield from the slow pyrolysis of *C. protothecoides* was temperature dependent and was 5.7%–55.3% [30].

Pyrolysis was found to produce less energy than it consumed, mainly due to the high energy penalty of the drying process of the wet microalgae, which contained around 80% of moisture. Despite the advantage of processing the whole microalgal biomass, this high energy cost is a significant hindrance that may prevent the complete development of this technology [31].

3.5. Properties

Besides high productivities and yields, fuel properties are important in the selection of the most adequate species for biofuel production. Table 2 gives some important fuel properties of two micralagal biodiesel samples.

TABLE 2. Algae biodiesel properties [38].

Parameter	Algae 1	Algae 2	Unit
Free-fatty acids	0.45	1.75	% weight
Cloud point	−5.2	3.9	°C
Cold filter plugging point	−7	2	°C
Free Glycerin	0.009	0.014	mass%
Total Glycerin	0.091	0.102	mass%
Monoglycerides	0.265	0.292	mass%
Diglycerides	0.078	0.070	mass%
Triglycerides	0.020	0.019	mass%
Water & sediment	<0.005	<0.005	% volume
Acid number	0.022	0.003	mg KOH/g
Visual inspection	1	1	Haze
Relative density at 60 F	0.8780	0.8780	N/A
Oxidative stability (110 °C)	8.5	11	h
Flash point (closed cup)	>160	>160	°C
Moisture	0.037	0.026	mass%
Cold soak filtration	85	84	s
Sulfur	5.1	0.6	ppm
Calcium	<0.1	0.7	ppm (μg/g)
Magnesium	0.3	1.1	ppm (μg/g)
Phosphorus	<0.1	<0.1	mass%
Carbon residue	0.007	0.042	mass%
Sulfated ash	<0.005	<0.005	mass%
Kinematic viscosity at 40 °C	4.519	4.624	mm^2/s
Copper corrosion (3 h at 50 °C)	1a	1a	N/A

3.6. Commercial Biofuel—Worldwide Production

Table 3 gives a breakdown of algal biofuel processing capacity due to come online until 2017. Algenol's first commercial plant, a 250,000 L plant based on algae fermentation is due to come online in 2016 in Mexico, significantly increasing the overall capacity. Solazyme are producing renewable oils based on hydroprocessing technology. In France, 15,000 L of capacity were due to come online in 2012 with another 27,000 L due to come online in Brazil in 2013.

TABLE 3. Projected algae biofuel capacity worldwide (million L) [39].

Technology	2011	2012	2013	2014	2015	2016	2017
Fermentation	0.38	0.38	0.38	0.38	0.38	945.38	945.38
Hydroprocessing	0.57	61.88	162.69	166.47	166.47	166.47	166.47
Transesterification	1.87	2.06	14.42	14.42	14.42	14.42	14.42
Other	0.38	0.76	0.76	1.66	1.66	1.66	1.66
Total	3.20	65.08	178.25	182.94	182.94	1127.94	1127.94

Figure 2 shows two pie charts representing the percentage distribution of algal biofuel producing companies around the world in different regions, and the percentage distribution of various algae production technologies being used by these companies [6]. These figures indicate that the majority of companies involved in algal biofuel development are based in the USA, with fewer European companies involved. The figures also show that closed systems (photobioreactors) are receiving more attention for cultivation than open pond or natural systems.

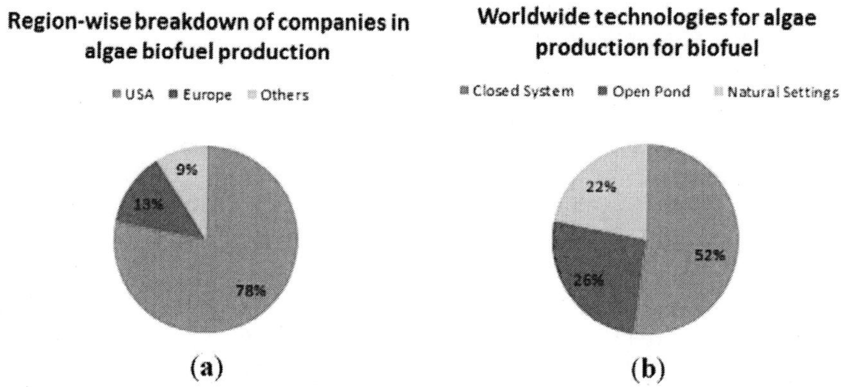

FIGURE 2. (a) Regions and (b) technologies in algae biofuel production, adapted from [6].

3.7. Resource Potential—Ireland

Assessing the potential availability and location of algal biomass is difficult due to uncertainty over the amount of land that is available and affordable in locations such as coastal areas, where the climate, water and nutrients may be sufficient to support the commercial cultivation of algae. Further research and development is required into the production of algal biofuels, as such any contribution to Irelands 2020 targets is likely to be minimal. Bruton [12] has estimated that approximately 79 TJ could come from microalgae resources by 2020 (approximately 2.3 million L).

The University of Limerick in Ireland is coordinating the EU-funded project "Direct Ethanol from MicroAlgae" (DEMA) [40]. The DEMA Consortium, composed of nine partners from both academia and industry from six EU countries, is aiming to develop, demonstrate and licence an economically competitive technology for the direct production of bioethanol from microalgae with low-cost scalable photobioreactors by 2016. The project is focusing on a strain of the cyanobacterium, *Synechocystis sp.*, which will be metabolically engineered to directly transform carbon dioxide, water and sunlight into bioethanol at a concentration level of >1%–2% (*v*/*v*). Subsequently, the bioethanol will be continuously extracted from the culture media via a membrane technology process exploiting existing EU expertise and technology. Initial proof-of-concept results from life cycle assessments and economic balances suggest the feasibility of using microalgae to produce bioethanol for less than 0.40/L.

4. MACROALGAE

Macroalgae are multicellular, macroscopic algae, which are abundant in coastal environments, primarily in nearshore coastal waters with suitable substrate for attachment. Such algae also occur as floating forms in the open ocean [41]. Macroalgae have relatively high photon conversion efficiency and can therefore rapidly synthesize biomass through assimilating abundant resources in nature such as sunlight, carbon dioxide, and inorganic nutrients [1]. The fast growing macroalgae are capable of yielding more biomass (*ca.* 2 kg C m^{-2} yr^{-1}) [4], than fast growing terrestrial crops such as sugarcane [42]. Macroalgae can grow to a considerable size (up to tens of meters in length in the case of Pacific kelp species) [43], although Atlantic species are smaller at approximately 4 m in length [42].

Macroalgae traditionally have not been considered as feedstocks for bioenergy production, but have been used in food, in medicine, or as fertilizer and in the processing of phycolloids and chemicals [8]. Research and development for a marine bioenergy and biofuel industry is still in its infancy in Ireland [44], however, macroalgae are already farmed on a large scale in Asia and to a lesser extent in Europe, primarily in France, and on a research scale in Scotland [8].

Macroalgae are suitable for biofuel production, however they do not generally contain triglycerides and are therefore being considered for the natural sugars and other carbohydrates they contain, making them suitable for biogas and ethanol production rather than biodiesel. Macroalgae has a negative lower heating value and high moisture content (*ca.* 80%–85%) as such the most appropriate method of processing for energy is fermentation by anaerobic digestion, to create biogas or ethanol [19].

Table 4 outlines the carbohydrate composition for some species of macroalgae. Based on the presence or lack of phytopigments other than chlorophyll, marine macroalgae are classified into three major classes: brown algae (Phaeophyceae), red algae (Rhodophyceae), and green algae (Chlorophyceae) [1]. Cell wall components of algae are the major sources of carbohydrates. Green algae mostly contain cellulose and hemicellulose; red algae contain cellulose and polysaccharides like agar and carrageenan; and brown algae contain cellulose and alginic acids.

There are approximately 1800 species of brown algae, and most are marine algae. Brown algae are generally the largest of the algae species, and are more often found in colder waters. The majority of algae biomass worldwide comes from a relatively small number of species in the orders Laminariales and Fucales [45]. The subtidal large brown kelps of the order Laminariales have been identified as having the greatest potential for bioconversion to energy [46]. Among green macroalgae, *Ulva* is the most important prospect from an energy perspective [12].

TABLE 4. Seaweed composition and sugars released by hydrolysis (% *w/w* dry biomass) for a variety of species [1].

Seaweed	Class	Carbohydrate composition	Total carbohydrates (%)	Lipid (%)	Protein (%)	Ash (%)	Sugars released by hydrolysis	Sugar composition
Gelidium amansii	Red	Agar, Carrageenan, Cellulose	75.2–83.6	0.6–1.1	12.2–18.5	3.3–5.7	34.6–67.5	Glucose, Galactose
Laminaria japonica	Brown	Laminarin, Mannitol, Alginate, Fucoidan, Cellulose	51.9–59.5	1.5–1.8	8.1–14.8	30.9–31.5	9.6–37.6	Glucose, Mannitol
Sargassum fulvellum			39.6	1.4	13	46	9.6	
Ulva lactuca	Green	Starch, Cellulose	54.3	6.2	20.6	18.9	19.4	Glucose
Ulva pertusa			65.2	2.6	7.0	25.2	59.6	

4.1. Cultivation and Harvest

There are essentially two ways of obtaining marine macroalgae; harvesting natural stocks or cultivation. Seaweed exploitation in Europe is currently restricted to manual and mechanized harvesting of natural stocks. The majority of Asian seaweed resources are cultivated [12]. Figure 3 gives a breakdown of wild harvest *versus*cultivated macroalgae along with the major countries involved.

Macroalgae for biofuel purposes are mainly harvested from the wild. Cultivation of seaweed does occur, mainly in Asian countries, but this is usually for higher value purposes such as food, or hydrocolooid (*i.e.*, agar-agar and alginate) production. Traditionally harvest is being done by hand, but this is now largely replaced by mechanical harvesting using trawler systems [47]. *Laminaria digitata* and *Laminaria hyperborea* are both harvested mechanically by boat in France and Norway. *Ascophyllum nodosum*, is harvested by boat in Norway. In France and Ireland, *Ascophyllum nodosum* is harvested manually. All the other species are harvested manually, either on foot or by diving [48].

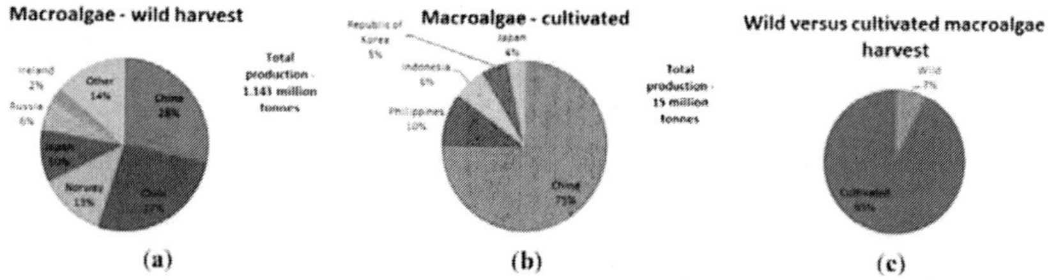

Figure 3. Breakdown of macroalgae (**a**) wild harvest by region; (**b**) cultivated harvest by region; and (**c**) wild *versus* cultivated in 2011 [49].

Macroalgae cultivation is currently in its infancy in Europe. Commercial aquaculture of seaweed is found in France (Brittany, six farms) Spain (Galicia, two farms) and on an experimental basis in Ireland, Asturias (Spain), Norway and the United Kingdom. The main cultivated species are *Saccharina latissima* and *Undaria pinnatifida*. In Ireland, *Palmaria palmata* farming is being experimented with on the west coast but the results seem limited.

In order to generate significant volumes of biomass for any biofuel industry, cultivation will be required in the long-term. Cultivation can occur, subject to appropriate licences, either at nearshore locations, or offshore. Over the last number of decades different cultivation systems for seaweeds have been developed and improved ranging from intertidal fixed and floating bottom farms for *Eucheuma*/*Kappaphycus* and *Gracilaria* (e.g., Philippines, Vietnam and Thailand) to elaborate floating net structures for *Porphyra* and long-line systems for kelp in China, Korea and Japan [50]. Modifications of long-line systems have been tested at small scale in Europe [8] amongst them a novel ring system from Germany [51]. These cultivation systems show that there is potential to develop large-scale ocean cultivation of seaweeds. However, existing cultivation and harvesting technology is labor intensive and needs to be optimized to reduce costs and energy demand [41,52].

Several designs and pilot systems have been developed for floating cultivation systems (farms) in open sea, some with surface areas up to 4000 hectares but no full-scale systems have been put in operation so far and their economic feasibility remains doubtful [3]. There is the possibility that cultivated macroalgae may be deployed through integration with existing aquaculture enterprises or in conjunction with new offshore wind farms [3]. Wind farms offer an ideal opportunity as the area is already closed for shipping, it would allow multifunctional use of area and offshore constructions, and joint operation and maintenance could occur involving synergy and cost benefits.

4.2. Processing

After cultivation and harvesting, the macroalgal biomass must be pre-treated for most biofuel applications. The first step of pre-treatment is to remove foreign objects and debris such as stones, sand, snails, or other litter that may be caught in the biomass either manually or by washing [12]. In many cases, chopping or milling is then required to increase the surface area/volume ratio [41]. Finally, the biomass should be dewatered to 20%–30% to increase shelf life and reduce transportation costs in situations where it must be stored for long periods or transported over long distances before further processing [1].

The principal energy process considered for seaweed is fermentation, either anaerobic digestion (AD), to create biogas, or ethanol fermentation. Other thermochemical options for macroalgae utilization include direct combustion, gasification, pyrolysis and liquefaction.

4.2.1. Fermentation

Hydrolysis of seaweeds converts the storage carbohydrate into simple fermentable sugars, which can be easily used to produce ethanol by some natural microorganisms [1]. Major sugars from brown seaweeds include glucan (laminarin or cellulose), mannitol, and alginates. However, natural microorganisms cannot utilize these various sugars concurrently, because they lack the ability to use alginate, and thus ethanol production from brown seaweed cannot reach its maximum level [1]. A recent breakthrough by Wargacki *et al.* [53] in the efficient use of brown seaweed biomass arose from an engineered *E. coli* platform capable of alginate degradation and metabolism. Here, ethanol was produced through co-fermentation of glucose, mannitol, and alginate from brown seaweed. This breakthrough allows a theoretical ethanol outcome of 80% conversion. As such 0.281 L of ethanol could be obtained per kg of dry seaweed.

4.2.2. Anaerobic Digestion

Macroalgae with low lipid content are especially suitable for biogas production using anaerobic fermentation. However, macroalgal biomass has some drawbacks over other types of biomass. For marine species, the high saline content may inhibit growth or productivity of the anaerobic microorganisms in the fermenter. This can be mitigated by mixing algal biomass with other types of biomass to "dilute" the saline concentration. Another problem, typical for green macroalgae, is the formation of H_2S

due to the high sulphate concentration in these species. This can be solved by using iron-based chemicals to bind H_2S, as has been applied in wastewater treatment systems [47]. Seaweeds have methane yields ranging from 0.14 m^3 kg^{-1} to 0.40 m^3 kg^{-1} volatile solids [41].

4.2.3. Hydrothermal Liquefaction

Hydrothermal liquefaction is a promising method of processing algae which takes advantage of the high moisture content of algae though reaching supercritical state [47], as discussed for microalgae above. Recently, a study on the hydrothermal liquefaction of the green seaweed *Enteromorpha prolifera* was made by Zhou *et al.* [54], which showed maximum yield of bio-crude of 23 wt% with an energy density of 29.89 MJ/kg.

Anastasakis and Ross [55] investigated the influence of reaction conditions on the liquefaction behavior of a typical brown seaweed commonly found around the British Isles, *Laminaria saccharina*. *L. saccharina* has high carbohydrate content and contains large amounts of mannitol and laminarin. The bio-crude yields ranged from 3.8 wt% to 19.3 wt%.

4.2.4. Pyrolysis

The conditions of pyrolysis are described previously for microalgae. The application of pyrolysis to macrolagae has been studied, however it suffers a number of serious problems which affects its viability; it requires relatively dry material and so is only viable when high-value products are formed [55], there are high costs associated with bulk handling and transport and there is difficulty in separating the complex mixtures of chemicals [8].

4.2.5. Transesterification

Transesterification is not a commonly investigated method for processing macroalgae due to their low lipid content, however transesterification has been used to produce biodiesel from the green macroalgae *Enteromorpha compressa* [56]. It was found that, by carrying out a base transesterification through optimum conditions of 1% NaOH, 9:1 methanol-oil ratio, 600 rpm and 60 °C temperature for 70 min, the maximum biodiesel yield of 90.6% of the oil used in the reaction was achieved [56].

4.3. Biorefinery Concept

Macroalgae contain a range of polysaccharides, proteins, minerals [3], and some species produce higher value compounds such as alginates, fucoidan, and mannitol [47]. As such, macroalgae is a suitable feedstock for biorefinery for co-production of chemicals and fuels to attain optimum valorization, as shown in Figure 4. Similar to how crude oil is refined in both fuel and fine chemicals; the value of algae is greatly increased if the parts of the biomass that cannot be converted into fuels are utilized for food, feed, chemicals, cosmetics, biomaterials or even pharmaceutical applications [57].

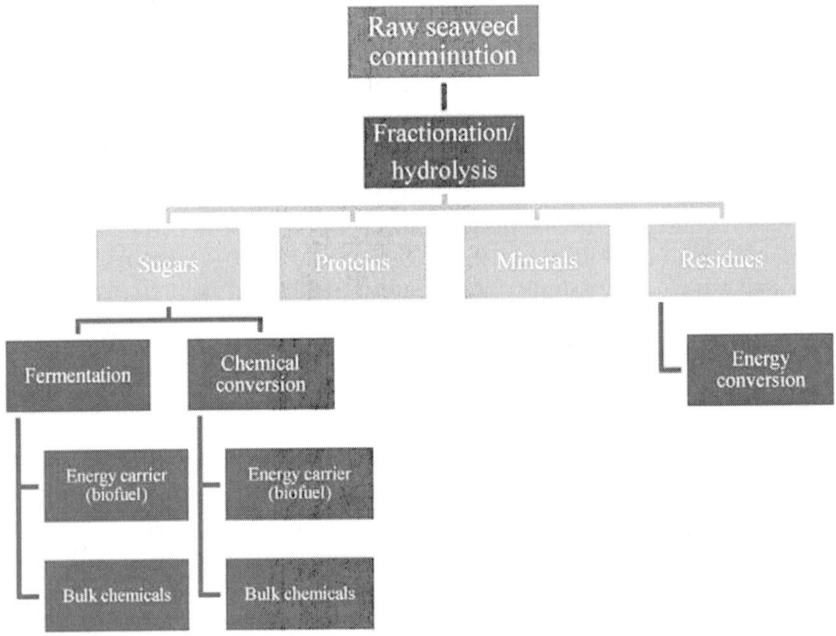

FIGURE 4. Macroalgae biorefinery concept, adapted from [3].

4.4. Properties

Table 5 displays some important fuel properties of macroalgal biodiesel in relation to the American Society for Testing and Materials (ASTM) fuel Standards.

4.5. Commercial Biofuel from Macroalgae

Biofuel from macroalgae is in its infancy. Statoil has entered into a partnership with Seaweed Energy Solutions AS (SES) and Bio Architecture Lab (BAL), to develop a macroalgae-to-ethanol system in

Norway. The aim of the partnerships is to develop a 10,000 ha seaweed farm off the coast of Norway which will produce 200,000 t of ethanol (2% of EU's ethanol marked) [58]. SES is developing the technology for large scale cultivation and harvesting technology [59], while BAL is responsible for developing the technology and process to convert the macroalgae into ethanol [60].

TABLE 5. Physicochemical characterization of algal biodiesel compared with American Society for Testing and Materials (ASTM) Standards (ASTM D 6751-06) [56].

Properties	Units	Algal biodiesel	ASTM limits
Flash point	°C	166	>160 min
Kinematic viscosity at 40 °C	$mm^2 s^{-1}$	4.35	1.9–6.0
Water and sediment	vol.%	0.005	0.050 max
Density at 15 °C	$kg m^{-3}$	878.47	n/a
Cetane number	-	58.5	47 min
Cloud point	°C	3	n/a
Acid value	$mg KOH g^{-1}$	0.43	0.8
Free glycerine	mass%	0.0034	0.020
Total glycerin	mass%	0.123	0.240
Sulfated ash	mass%	0.0024	0.020
Pour point	°C	−2	n/a
Carbon residue	mass%	0.01	0.050
Sulfur	mass%	0.00056	0.05
Copper strip corrosion	-	1	No. 3 max
Distillation temperature	°C	346	360
Phosphorous	mass%	0.0004	0.001

4.6. Macroalgae in Ireland

Ireland is one of the largest producers of macroalgae in Europe, producing 29,500 t, 13% of the European total (226,500 t. Approximately 30,000 t of algae is processed in Ireland annually [61]. Location of macroalgae processing capacity in Irland is shown in Figure 5. Currently the majority of seaweed collection in Ireland is for human consumption and for hydrocolloid production, with 29,000 t harvested in 2006 for Arramara Teo, a state owned company [19]. The *Ascophyllum nodosum* species dominates the industry, accounting for approximately 25,000 t or 95% of domestic production. *Ascophyllum nodosum* is processed at two factories on the west coast (Donegal & Galway) and is used to produce fertilizers, horticultural products and animal feed, as shown inFigure 5 and Figure 6. A significant quantity of national production is sold as raw material for further industrial processing.

Numerous other species are harvested and used for commercial purposes in Ireland including; *Fucusserratus, Chondrus crispus, Laminaria digitata, Fucus vesiculosus* and *Saccharina latissima*. Ireland has been importing significant quantities of *Lithothamnion corallioides* from Iceland for processing into agricultural and nutritional products. Aquaculture of macroalgae is still largely experimental in Ireland and has not contributed significantly to domestic production of algae, experimental cultivation of *Asparagopsis armata, Alaria esculenta, Palmaria palmata,Laminaria digitata* and *Porphyra* has been achieved over the last 20 years [61].

Ireland has considerable macroalgal resources, with an estimation of three million t of standing kelp stock, however this estimation is highly uncertain [19]. Barriers exist to the further exploitation of marine biomass, in particular any potential effects on marine biodiversity [19].

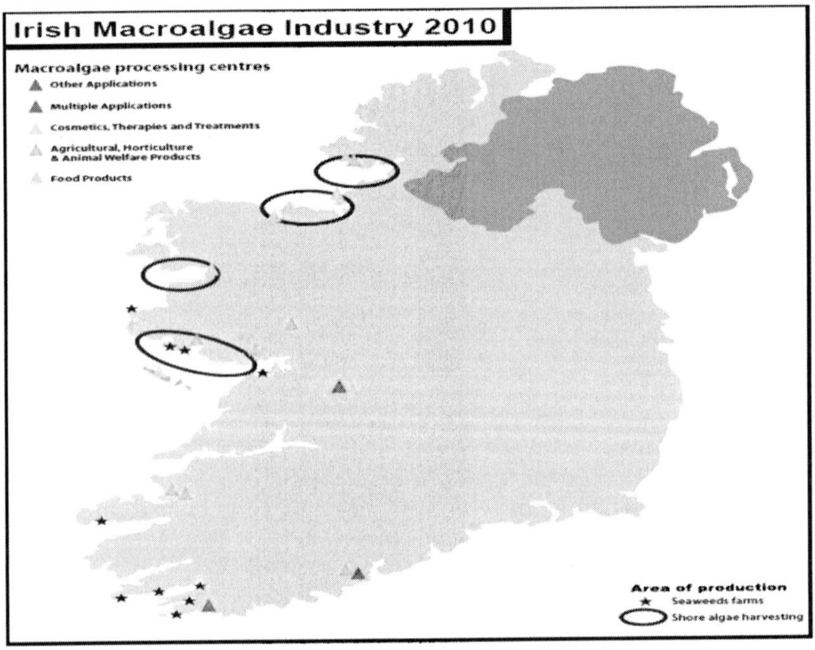

FIGURE 5. Irish macroalgae industry 2010 [61].

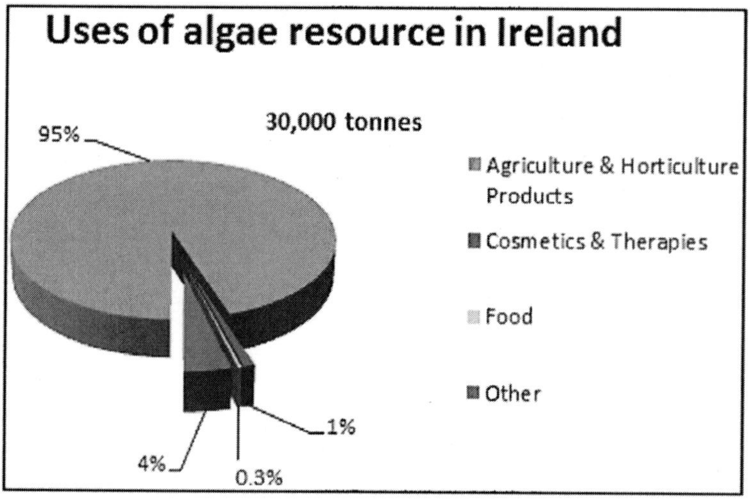

FIGURE 6. Uses of macroalgae in Ireland [61].

Recently, the potential of algae for bio-energy production and strong interest in developing integrated multi-trophic aquaculture systems has given a new dimension to algae aquaculture [61].

The BioMara project was a joint UK and Irish project that aimed to demonstrate the feasibility and viability of producing third generation biofuels from marine biomass [62]. BioMara was a partnership between six scientific institutes in the EU Interreg IVA cross-border collaboration area (the Border region of Ireland and Northern Ireland, and Western Scotland). Dundalk Institute of Technology and Sligo Institute of Technology were the two Irish partners involved. As part of the project five potential seaweed species common in Irish waters, *Saccorhiza polyschides*, *Ulva sp.*, *Laminaria digitata*, *Fucus serratus* and *Saccharina latissima*, were co-digested individually with bovine slurry [63]. The combination of *S. latissima* with slurry produced the maximum methane yield (335 mL g volatile solids^{-1}) followed by *S. polyschides* with (255 mL g volatile solids^{-1}). The methane and CO_2 percentages ranged between 50%–72% and 10%–45%, respectively. The results of the study established that the seaweed species investigated are promising feedstocks for the production of biogas and methane for energy production. However, their use on a large-scale will require further development to increase yields and reduce production costs [63].

Sanderson [8] estimated the costs of setting up a one hectare seaweed farm in the UK. The study was based on a detailed breakdown of the costs per 40 m longline (surface barrelfloats, headline, anchor rope, stainless steel swivels, thimbles for splicing, shackles and mooring blocks) of £2382.49, depreciated over five years, together with approximations of other annual costs. It was estimated that the one hectare plot would need to yield 130 t wet weight, worth £500/t to break even [64]. However, the study notes that the estimated costs are potentially over estimates, and could be considerably lower given economies of scale. For example, there is the opportunity to avail of technology transfer from countries where macroalgae are cultivated on a very large scale, as in areas in Asia. Reducing labor costs in the UK could be achieved by further mechanization of the culture methods and would be vital in improving the economic viability of the system [8].

In Ireland, the Mabfuel project at the Daithi O'Mhurchu Marine Research Station [65] is just reaching its conclusion after three years of investigating the feasibility of using algae as a feedstock for producing bio-fuels in Ireland and Turkey. The Mabfuel project aimed to develop optimal methods to extract oil from algal biomass and to develop intensive large-scale culture methods for micro-algal species for indoor and outdoor facilities. The project was run in conjunction with seven partners, including two Turkish universities as well as Green Biofuels Ireland, which operates a biofuel plant at New Ross, Ireland. A business plan is being prepared with the aim of becoming Ireland's first commercial scale producer of algae for biofuels.

Energetic Algae ("EnAlgae") [66], is a multi-year project (2009–2015) aiming to investigate the potential of reducing GHG emissions in the North West Europe (NWE) region by producing microalgae and macroalgae for biofuel. The project brings together 19 partners and 14 observers across 7 EU Member States and has received funding from the European Regional Development Fund through Interreg IVB [67]. The National University of Ireland Galway (NUIG) is involved in work package one of the project, the objective of which is to maximise the transnational value of pilot scale algal culture facilities across NWE. NUIG has developed a large-scale hatchery which is used to grow kelps, including *Saccharina latissima* ("Sugar kelp") and *Alaria esculenta* ("Atlantic wakame"). The aim of this facility is to standardise and increase the production of a hectare site at sea. It is hoped that by monitoring of the environmental conditions of the site, considerable insights may be gained into macroalgal yield is influenced by surrounding habitat.

University College Dublin (UCD), is involved in work packages two and three of the project which aim to "identify political, economic, social and technological opportunities which promote the adoption of algal biomass within NEW" and "to combine information across the algal bioenergy delivery chain into a comprehensive and user friendly ICT tool which will facilitate decision making, identify gaps in current knowledge and capability, and provide a roadmap by which stakeholders can focus future actions in the region" [68]. UCD is working to develop materials and methods for the sustainable conversion of biomass to bioenergy, and ways is which this energy can be stored. UCD studies on artificial photosynthesis, the production of H_2 and hydrocarbon from water and CO_2 [69], will assist the EnAlgae project.

Further research and development is required into the production of algal biofuels, as such any contribution to Ireland's 2020 targets is likely to be minimal. Bruton [19] has that up to 447 TJ (approximately 21 million L) of energy may be generated from macroalgae by 2020.

5. LIFE CYCLE ASSESSMENT (LCA) OF MICRO AND MACROALGAE SYSTEMS

From 2013 to 2017, biofuels must achieve a minimum 35% reduction in GHG emissions *versus* fossil fuels.

5.1. Microalgae

Slade and Bauen [14] reviewed reported energy balances and associated GHG emissions from a number of studies of microalgae production systems. They found that the net energy ratio (the sum of the energy used for cultivation, harvesting and drying, divided by the energy content of the dry biomass) is less than one for 75% of the open pond studies reported. In these systems there are high energy requirements associated with pumping, fertilizer production, and drying and de-watering. However, the net energy ratios of all photobioreactor studies reported are all greater than one, with the majority of energy requirements associated with cultivation and harvesting. The results show that while positive energy balances are possible, algae production systems will need to be optimized and technical advances will be required to make these systems viable.

The associated carbon dioxide emissions associated with the algae production systems were estimated by multiplying the external energy inputs reported in the studies by the default emissions factors described in the EU renewable energy directive [14]. The production of algae biomass by the open pond method is comparable in terms of CO_2 emissions to the production of rape methyl ester biodisesel (45 g CO_2 eq. MJ^{-1}), while production in photobioreactors produces emissions higher than conventional fossil diesel (84 g CO_2 eq. MJ^{-1}) [14].

5.2. Macroalgae

Alvarado-Morales *et al.* [70] analyzed the energy requirements and environmental impacts of two macroalgae-based biofuel scenarios in Scandinavia. The two scenarios analyzed were biogas production (Scenario 1), and bioethanol with biogas production (Scenario 2), from the brown seaweed (*Laminaria digitata*). The production of seaweed was identified to be the most energy intensive step, mainly due to the energy required in the land based growth of juveniles and the grow-out phase. The net energy consumption was 2.26 and 3.04 GJ/t of dry macroalgae for Scenarios 1 and 2, respectively. The energy produced from Scenario 1 was approximately 555 kW h (2 GJ) of electricity per tonne of dried macroalgae, resulting in a negative energy balance for the system. For Scenario 2, the energy produced amounted to 2.01 GJ/75 kg of produced ethanol and 359 kW h (1.29 GJ) of electricity per tonne of dry macroalgae. This corresponds to a total energy production of 3.30 GJ/t of dry macroalgae, therefore resulting in a slight positive energy return. In terms of GHG emissions, Scenario 1 has lower emissions compared to Scenario 2, where the increase in emissions can be attributed to energy consumption for bioethanol production downstream and the purification process [70]. The study highlights the need for development of seaweed production, bioethanol distillation, and management of digestate in order to improve the environmental performance of macraolgae biofuel production.

6. CONCLUSIONS

While the algae industry is relatively well developed for food, feed and nutraceutical applications, along with high value products, algae for biofuel is in its infancy. The industry still requires considerable research and development in order to develop economically viable algae biofuel production. The high biomass productivity of both micro and macroalgae, along with favorable biomass compositions, make them ideal feedstocks for conversion into a range of biofuels. Microalgae is more typically suited to biodiesel production due to its high lipid content, while the high levels of natural sugars in macroalgae makes it more suitable for fermentation to produce ethanol. The technology for production and conversion of microalgae to biofuels has not reached any appreciable commercial scale, with only one or two algae biofuel plants coming online in recent years. Macroalgal biofuel is limited to a research and demonstration scale.

It is predicted that microalgal biofuel capacity will reach 1128 million L by 2017. The majority of companies involved in algae production are based in the USA, this could influence trade flows. The costs for aquatic biomass production for energy applications are currently too high to compete with bio energy applications from biomass produced on land. However, there is potential to improve the value of algae if the parts of the biomass that cannot be converted into fuels are utilized for food, feed, chemicals, cosmetics, biomaterials or even pharmaceutical applications.

ACKNOWLEDGMENTS

This study was funded under the Charles Parsons Energy Research Program (Grant Number 6C/CP/E001) of Science Foundation Ireland (SFI).

REFERENCES

1. Wei, N.; Quarterman, J.; Jin, Y.-S. Marine macroalgae: An untapped resource for producing fuels and chemicals. *Trends Biotechnol.* **2013**, *31*, 70–77.
2. Schenk, P.M.; Thomas-Hall, S.R.; Stephens, E.; Marx, U.C.; Mussgnug, J.H.; Posten, C.; Kruse, O.; Hankamer, B. Second generation biofuels: High-efficiency microalgae for biodiesel production. *Bioenergy Res.* **2008**, *1*, 20–43.
3. Reith, J.H.; van Hal, J.W.; Lenstra, W.J. Large-Scale Carbon Recycling via Cultivation and Biorefinery of Seaweeds for Production of Biobased Chemicals and Fuels. In Proceedings of the Conference on Carbon Dioxide as Feedstock for Chemistry and Polymers, Essen, Germany, 10–11 October 2012.
4. Skjermo, J. *Macroalgae Activities in Norway*; Nordic Algae Workshop: Grenaa, Denmark, 2012.

5. Daroch, M.; Geng, S.; Wang, G. Recent advances in liquid biofuel production from algal feedstocks. *Appl. Energy* **2013**, *102*, 1371–1381.
6. Singh, J.; Gu, S. Commercialization potential of microalgae for biofuels production. *Renew. Sustain. Energy Rev.* **2010**, *14*, 2596–2610.
7. Adams, J.M.M.; Ross, A.B.; Anastasakis, K.; Hodgson, E.M.; Gallagher, J.A.; Jones, J.M.; Donnison, I.S. Seasonal variation in the chemical composition of the bioenergy feedstock *Laminaria digitata* for thermochemical conversion. *Bioresour. Technol.* **2011**, *102*, 226–234
8. Kelly, M.S.; Dworjanyn, S. *The Potential of Marine Biomass for Anaerobic Biogas Production: A Feasibility Study with Recommendations for Further Research*; Marine Estate Research Report; The Crown Estate, the Scottish Association for Marine Science: Scotland, UK, 2008.
9. European Commission. Directive 2009/28/EC of the European Parliament and of the Council of 23 April 2009 on the promotion of the use of energy from renewable sources and amending and subsequently repealing Directives 2001/77/EC and 2003/30/EC. *Off. J. Eur. Union* **2009**, *L40*, 16–47.
10. Benemann, J.R.; Oswald, W.J. *Systems and Economic Analysis of Microalgae Ponds for Conversion of CO_2 to Biomass*; University of California: Berkeley, CA, USA, 1996.
11. Singh, A.; Olsen, S.I. A critical review of biochemical conversion, sustainability and life cycle assessment of algal biofuels. *Appl. Energy* **2011**, *88*, 3548–3555.
12. Bruton, T.; Lyons, H.; Lerat, Y.; Stanley, M.; Rasmussen, M.B. *A Review of the Potential of Marine Algae as a Source of Biofuel in Ireland*; Sustainable Energy Ireland: Dublin, Ireland, 2009.
13. Singh, A.; Nigam, P.S.; Murphy, J.D. Mechanism and challenges in commercialisation of algal biofuels.*Bioresour. Technol.* **2011**, *102*, 26–34.
14. Slade, R.; Bauen, A. Micro-algae cultivation for biofuels: Cost, energy balance, environmental impacts and future prospects. *Biomass Bioenergy* **2013**, *53*, 29–38.
15. Kanda, H.; Li, P. Simple extraction method of green crude from natural blue-green microalgae by dimethyl ether. *Fuel* **2011**, *90*, 1264–1266.

16. Mata, T.M.; Martins, A.A.; Caetano, N.S. Microalgae for biodiesel production and other applications: A review. *Renew. Sustain. Energy Rev.* **2010**, *14*, 217–232.
17. Ravishankar, G.A.; Sarada, R.; Vidyashankar, S.; VenuGopal, K.S.; Kumudha, A. Cultivation of Micro-Algae for Lipids and Hydrocarbons, and Utilization of Spent Biomass for Livestock Feed and for Bio-Active Constituents. In *Biofuel Co-Products for Livestock Feed—Opportunities and Challenges*; Makkar, H.P.S., Ed.; Food and Agriculture Organization of the United Nations (FAO): Rome, Italy, 2012; Chapter 24; pp. 423–446.
18. Singh, N.K.; Dhar, D.W. Microalgae as second generation biofuel. A review. *Agron. Sustain. Dev.* **2011**, *31*, 605–629.
19. Lam, M.K.; Lee, K.T. Microalgae biofuels: A critical review of issues, problems and the way forward.*Biotechnol. Adv.* **2012**, *30*, 673–690.
20. Bezergianni, S.; Dimitriadis, A.; Kalogianni, A.; Pilavachi, P.A. Hydrotreating of waste cooking oil for biodiesel production. Part I: Effect of temperature on product yields and heteroatom removal. *Bioresour. Technol.* **2010**, *101*, 6651–6656.
21. Kanes, S.; Forster, D. *The Choice of Next-Generation Biofuels (Algae Excerpt)*; Equity Research Industry Report; Scotia Capital: New York, NY, USA, 2009.
22. Ghasemi, Y.; Rasoul-Amini, S.; Naseri, A.T.; Montazeri-Najafabady, N.; Mobasher, M.A.; Dabbagh, F. Microalgae biofuel potentials (Review). *Appl. Biochem. Microbiol.* **2012**, *48*, 126–144.
23. *Jet Fuel from Microalgal Lipids*; National Renewable Energy Laboratory: Golden, CO, USA, 2006.
24. Guzman, A.; Torres, J.E.; Prada, L.P.; Nuñez, M.L. Hydroprocessing of crude palm oil at pilot plant scale.*Catal. Today* **2010**, *156*, 38–43.
25. Chernova, N.I.; Korobkova, T.P.; Kiseleva, S.V. Use of biomass for producing liquid fuel: Current state and innovations. *Therm. Eng.* **2010**, *57*, 937–945.
26. Toor, S.S.; Rosendahl, L.; Rudolf, A. Hydrothermal liquefaction of biomass: A review of subcritical water technologies. *Energy* **2011**, *36*, 2328–2342.
27. Brown, T.M.; Duan, P.; Savage, P.E. Hydrothermal liquefaction and gasification of *Nannochloropsis* sp.*Energy Fuels* **2010**, *24*, 3639–3646.

28. Biller, P.; Ross, A.B. Potential yields and properties of oil from the hydrothermal liquefaction of microalgae with different biochemical content. *Bioresour. Technol.* **2011**, *102*, 215–225.
29. Jena, U.; Das, K.C.; Kastner, J.R. Effect of operating conditions of thermochemical liquefaction on biocrude production from *Spirulina platensis*. *Bioresour. Technol.* **2011**, *102*, 6221–6229.
30. Jena, U.; Das, K.C. Comparative evaluation of thermochemical liquefaction and pyrolysis for bio-oil production from microalgae. *Energy Fuels* **2011**, *25*, 5472–5482.
31. López Barreiro, D.; Prins, W.; Ronsse, F.; Brilman, W. Hydrothermal liquefaction (HTL) of microalgae for biofuel production: State of the art review and future prospects. *Biomass Bioenergy* **2013**, *53*, 113–127.
32. Biller, P.; Riley, R.; Ross, A.B. Catalytic hydrothermal processing of microalgae: Decomposition and upgrading of lipids. *Bioresour. Technol.* **2011**, *102*, 4841–4848.
33. Suali, E.; Sarbatly, R. Conversion of microalgae to biofuel. *Renew. Sustain. Energy Rev.* **2012**, *16*, 4316–4342.
34. Hirano, A.; Hon-Nami, K.; Kunito, S.; Hada, M.; Ogushi, Y. Temperature effect on continuous gasification of microalgal biomass: Theoretical yield of methanol production and its energy balance. *Catal. Today* **1998**,*45*, 399–404.
35. Sánchez Mirón, A.; Cerón García, M.-C.; García Camacho, F.; Molina Grima, E.; Chisti, Y. Growth and biochemical characterization of microalgal biomass produced in bubble column and airlift photobioreactors: Studies in fed-batch culture. *Enzym. Microb. Technol.* **2002**, *31*, 1015–1023.
36. Miao, X.; Wu, Q.; Yang, C. Fast pyrolysis of microalgae to produce renewable fuels. *J. Anal. Appl. Pyrolysis***2004**, *71*, 855–863.
37. Demirbaş, A. Oily products from mosses and algae via pyrolysis. *Energy Sources Part A Recover. Util. Environ. Eff.* **2006**, *28*, 933–940.
38. Sanford, S.D.; White, J.M.; Shah, P.S.; Wee, C.; Valverde, M.A.; Meier, G.R. *Feedstock and Biodiesel Characteristics Report*; Renewable Energy Group, Inc.: Ames, IA, USA, 2009.]
39. Biofuels Digest Advanced Biofuels & Chemicals Project Database Release Q3 2012 (07/27/12-Revision 1.1). Available

online: http://www. biofuelsdigest. com/ bdigest/wp-content/ uploads / 2012/07/ ABPD- Q312.xls(accessed on 20 March 2013).
40. European Commission: Community Research and Development Information Service (CORDIS): Projects: Direct Ethanol from MicroAlgae (DEMA). Available online: http://cordis.europa.eu/projects/rcn/106280_en.html (accessed on 9 October 2013).
41. Roesijadi, G.; Jones, S.B.; Snowden-Swan, L.J.; Zhu, Y. *Macroalgae as a Biomass Feedstock: A Preliminary Analysis*; Pacific Northwest National Laboratory: Richland, WA, USA, 2010.
42. Adams, J.M.M.; Toop, T.A.; Donnison, I.S.; Gallagher, J.A. Seasonal variation in *Laminaria digitata* and its impact on biochemical conversion routes to biofuels. *Bioresour. Technol.* **2011**, *102*, 9976–9984.
43. Kraan, S. Mass-cultivation of carbohydrate rich macroalgae, a possible solution for sustainable biofuel production. *Mitig. Adapt. Strateg. Glob. Chang.* **2013**, *18*, 27–46.
44. Roberts, T.; Upham, P. Prospects for the use of macro-algae for fuel in Ireland and the UK: An overview of marine management issues. *Mar. Policy* **2012**, *36*, 1047–1053.
45. The Seaweed Site: Information on Marine Algae. Available online: http://www.seaweed.ie/algae/phaeophyta.php (accessed on 11 March 2013).
46. Hughes, A.D.; Black, K.D.; Campbell, I.; Heymans, J.J.; Orr, K.K.; Stanley, M.S.; Kelly, M.S. Comments on "Prospects for the use of macroalgae for fuel in Ireland and UK: An overview of marine management issues". *Mar. Policy* **2013**, *38*, 554–556.
47. Streefland, M.; Rous, J.-F.; Zachleder, V.; Casteleyn, G.; Vernon, P.-A.; Acien, G.F.; Garofalo, R. *Report on Biofuel Production Processes from Micro-, Microalgae and Other Aquatic Biomass*; AquaFUELs: Brussels, Belgium, 2010.
48. *Seaweed Industry in Europe*; Netalgae: Cork, Ireland, 2012.
49. *Fishery and Aquaculture Statistics—Aquaculture Production*; Food and Agriculture Organisation of the United Nations (FAO): Rome, Italy, 2010.
50. Critchley, A.T.; Ohno, M.; Largo, D.B. World seaweed resources. An authoritative reference system. *Bot. Mar.* **2006**, *49*, 456–457.

51. Buck, B.H.; Buchholz, C.M. The offshore-ring: A new system design for the open ocean aquaculture of macroalgae. *J. Appl. Phycol.* **2004**, *16*, 355–368.
52. Troell, M.; Joyce, A.; Chopin, T.; Neori, A.; Buschmann, A.H.; Fang, J.-G. Ecological engineering in aquaculture—Potential for integrated multi-trophic aquaculture (IMTA) in marine offshore systems.*Aquaculture* **2009**, *297*, 1–9.
53. Wargacki, A.J.; Leonard, E.; Win, M.N.; Regitsky, D.D.; Santos, C.N.S.; Kim, P.B.; Cooper, S.R.; Raisner, R.M.; Herman, A.; Sivitz, A.B.; *et al.* An Engineered microbial platform for direct biofuel production from brown macroalgae. *Science* **2012**, *335*, 308–313.
54. Zhou, D.; Zhang, L.; Zhang, S.; Fu, H.; Chen, J. Hydrothermal liquefaction of macroalgae *Enteromorpha prolifera* to bio-oil. *Energy Fuels* **2010**, *24*, 4054–4061.
55. Anastasakis, K.; Ross, A.B. Hydrothermal liquefaction of the brown macro-alga *Laminaria Saccharina*: Effect of reaction conditions on product distribution and composition. *Bioresour. Technol.* **2011**, *102*, 4876–4883.
56. Suganya, T.; Nagendra Gandhi, N.; Renganathan, S. Production of algal biodiesel from marine macroalgae*Enteromorpha compressa* by two step process: Optimization and kinetic study. *Bioresour. Technol.* **2013**, *128*, 392–400.
57. Streefland, M.; Wiffels, R.; van Es, M.; Boussiba, S.; Leu, S.; Zachleder, V.; Verween, A.; Vieira, V.V.; Acien, G.F.; Molina-Grima, E.; *et al. Technological Assessment Including Downstream AddedValue Products*; AquaFUELs: Brussels, Belgium, 2011.
58. Ystanes, L.; Waage Fougner, M. *Seaweed to Biofuels—Future Perspectives by Industry Actor*; Marine Agronomy Group: Hilo, HI, USA, 2012.
59. *Seaweed Energy Solutions (SES) Awarded Contract by Statoil*; Seaweed Energy Solutions AS: Trondheim, Norway, 2012.
60. Statoil & Bio Architecture Lab Partner for Macroalgae-to-Ethanol. Available online: http://www.algaeindustrymagazine.com/statoil-bio-architecture-lab-partner-for-macroalgae-to-ethanol/ (accessed on 6 March 2013).
61. *Irish Macroalgae Industry*; Netalgae: County Cork, Ireland, 2011.
62. Welcome to the BioMara Project—BioMara. Available online: http://www.biomara.org/ (accessed on 11 October 2013).

63. Vanegas, C.H.; Bartlett, J. Green energy from marine algae: Biogas production and composition from the anaerobic digestion of Irish seaweed species. *Environ. Technol.* **2013**, *34*, 2277–2283.
64. Sanderson, J.C. Reducing the Environmental Impact of Seacage Fish Farming through Cultivation of Seaweed. Ph.D. Thesis, The Open University, Buckinghamshire, UK, 2006.
65. O'Sullivan, G.; McDonagh, N.; Pedreschi, D. *Irish Participation in EU FP7 Funded Competitive Marine Research Projects during the Period 2007–2008*; Marine Institute: Galway, Ireland, 2009. INTERREG IVB NWE Programme—Investing in Opportunities. Available online: http://www.nweurope.eu/index.php?act=project_detail&id=5376 (accessed on 9 October 2013).
66. EnAlgae Home Website. Available online: http://www.enalgae.eu/index.htm (accessed on 11 October 2013).
67. EnAlgae. EnAlgae Work Packages. Available online: http://www.enalgae.eu/enalgae-work-packages.htm (accessed on 11 October 2013).
68. Farrelly, D.; Fagan, C.; McDonnell, K. Effect of Carbon Dioxide Concentrations on the Carbon Mitigation Potential of Microalgae. In *Biosystems Engineering Research Review 17*; Cummins, E., Curran, T., Eds.; University College Dublin: Dublin, Ireland, 2012; pp. 123–126.
69. Alvarado-Morales, M.; Boldrin, A.; Karakashev, D.B.; Holdt, S.L.; Angelidaki, I.; Astrup, T. Life cycle assessment of biofuel production from brown seaweed in Nordic conditions. *Bioresour. Technol.* **2013**, *129*, 92–99.

11

CHAPTER

AVAILABLE RESOURCES FOR ALGAL BIOFUEL DEVELOPMENT IN CHINA

Shuhao Huo[1,2], Renjie Dong[1], Zhongming Wang[2,*], Changle Pang[1,*], Zhenhong Yuan[2], Shunni Zhu[2] and Li Chen[1]

[1]*College of Engineering, China Agricultural University, Beijing 100083, China*

[2]*Guangzhou Institute of Energy Conversion, Chinese Academy of Sciences, Guangzhou 510640, China*

ABSTRACT

Microalgal biofuel research in China has made noticeable progress, and algae cultivation for biofuel production is considered to be an important contribution to Greenhouse Gas (GHG) mitigation and energy security. In this paper, the algal biofuel potentiality in China was reviewed from the points of view of algal biodiversity, algal culture collection, GHGs (especially CO2) mitigation, and the availability of the required sunlight, wastewater and land resources. The cultivation of microalgae utilizing power plants gas with large amounts of CO2 and wastewaters from urban households, industry and animal husbandry are suitable for large scale production in China. Land is hardly a limitation for algae cultivation.

Keywords: algae; biofuel; biodiversity; resources; China

1. INTRODUCTION

Due to the concerns about energy security and climate change, biofuel development and utilization have received high attention in China. However, biofuel production from cereals, sugar crops, or animal fats, which is much easier and almost cost competitive with other biomass feedstocks, is not encouraged in China because it competes directly with food supplies. Lignocellulosic ethanol technology is still under development. The yield of biodiesel production is quite low compared with the need [1].

The promise of clean and sustainable energy production from algae has generated tremendous interest recently. Algae are the fastest-growing plants and the most widely distributed organisms in Nature. The dry cell weight contains about 50% oil [2]. Algae are a large group of photoautotrophic, mixotrophic and heterotrophic lower plants. Algae can be typically subdivided into two major categories, based on their relative sizes: macroalgae and microalgae. Microalgae are capable of producing 30 times the amount of oil per unit area of land, compared to oilseed crops [3]. It is the first largest biomass resource. Microalgae are abundant in diverse ecological aquatic habitats such as fresh, brackish, marine, and hypersaline water environments within a wide range of temperatures, pH values and salinity. Theoretically, microalgae could grow in different kind of water, nutrients, and land saving sites for oil production. Even better production can be achieved in the desert, where the sunlight is stronger [4].

There are a large number of potential pathways for the conversion of microalgal biomass into lipids, ethanol, hydrogen, methane, hydrocarbon and carbohydrates. The lipids can be refined into biodiesel for land transportation and even for aviation use. Microalgal biodiesel has become one of the potential third-generation biofuels. CO_2, sunlight, water, and land are the main natural resources required for algal production. For large-scale cultivation of microalgae, the major problem is how to reduce the cost of media. China has a lot of waste resources that can be utilized for biomass production. The recycling of waste resources is a win-win way for simultaneous industrial microalgae cultivation and waste treatment.

2. ALGAL BIODIVERSITY IN CHINA

With a large sea area, abundant freshwater resources, complex topography and diverse climates, China is one of the countries with the greatest diversity of algal resources [5]. Algae are distributed widely in oceans, various inland waters and wetland surfaces. Among them, those which live in the sea, oceans and inland saltwaters are saltwater algae, those which live in inland freshwater bodies are freshwater algae (Table 1).

TABLE 1. Comparison of the species number of algae between china and the world [5,6].

Category	In the World		In China		Freshwater (%)	Seawater (%)
	Freshwater	Seawater	Freshwater	Seawater		
Cyanobateria	1600	2000	1000	99	63	5
Euglenophyta	800	0	600	0	75	0
Diatomeae	8000	8336	4000	1485	50	18
Chrysophyta	701	49	100	17	14	35
Xanthophyta	381	20	25	5	7	25
Cryptophyta	60	19	15	6	25	32
Prymnesiophyta	0	500	0	34	0	7
Phaeophyta	8	1500	3	260	38	17
Dinozoa	50	4000	30	302	60	8
Rhodophyta	0	4100	0	569	0	14
Chlorophyta	8000	600	3000	163	38	27
Total	19,600	21,124	8773	2940	45	14

2.1. Algal Biodiversity in China Seas

In 2007, the 2940 species of algae recorded from Chinese seas belong to 10 phyla (calculated to represent about 14% of the total World species). In contrast, the number of species of algae recorded was 2589 in 1994. In recent decades, with the deepening of the research work, more species have been found, but species of some phyla like Cyanobateria and Chlorophyta are decreasing because of deteriorating water quality. Among them, the species of Diatomeae is the most dominant phylum and it takes up 40% of the World's marine primary productivity [7]. In 2007, the species of Diatomeae in China seas was 1485 or 50.5% of the total number of species.

Approximately 120 species of phytoplankton, dominated by eurythermal, low salinity, diatom species, have been recorded in the Bohai Sea. The dominant species are *Coscinodiscus, Chaetoceros, Rhizosolenia,* and *Skeletonema costatum*. While some green algae are dominant in the inter-tidal zones, brown and red algae dominant in sub-tidal zones have been recorded. In the Yellow Sea, about 386 species of phytoplankton, dominated by*Coscinodiscus, Chaetoceros, Rhizosolenia, Biddulphia, Nitzschia and*

Peridinium, have been recorded, with various ecologies indicative of temperate seas [5,6]. In the East China Sea, 64 species of phytoplankton occur in the areas near the estuary of the Yangtze River and 261 species appear in off-shore areas in Zhejiang. The dominant algae are *Coscinodiscus*, *Skeletonema costatum*, *Chaetoceros lorenzianus* and *Nitzschia pungens*. Neritic species are the most common ecological type. In the South China Sea, the number of phytoplankton species is different between regions and ranges from 104 to 260. They belong to two large groups, diatoms and dinoflagellates, among which*Chaetoceros* and *Rhizosolenia* are prevalent [5,6]. Since the 1990s, from the eastern Liaoning Bay to western Guangdong, more than half of the coastal sea water has been seriously polluted. Compared with category 2 in the Sea Water Quality Standard (GB3097-1997, GB means national standard), the inorganic nitrogen of coastal area of Yellow Sea and Bohai Sea has exceeded 46% and 45% and the heavy metals exceed 19% and 42%, respectively. The chemical oxygen demand of inshore Bohai Sea water exceeded 13%, which can cause serious damage to the marine algal biodiversity [8].

2.2. Algal Biodiversity in China Freshwater Bodies

There are diverse and abundant freshwater algae in China. The species of Diatomeae and Chlorophyta are more abundant than those of other phyla. The species number of freshwater algae in China accounts for 45% of the whole World number. Nearly a century of investigation has shown that all the categories of freshwater algae occur in China and the species are abundant. It has been shown that the total number of freshwater algae is about 25,000 and 9000 species have been recorded from China [5]. However, there are still many areas have not been surveyed and some areas have been surveyed, but not comprehensively.

Meanwhile, although the algal resources of China are plentiful, they are severely threatened. Nowadays more than 850 out of 1200 rivers in China have been contaminated, leading to the extinction of many species. The tasks of restoration and reconstruction of aquatic habitat are urgent [8] and the attention of conservationists to biodiversity only focuses on large animals and higher plants, but the small algae are generally neglected [5]. Among the largest groups of microalgae, Cyanophyceae, Chlorophyceae, Bacillariophyceae and Chrysophyceae are the most frequently cited as possessing the most desirable features for efficient and economical combination of CO_2 fixation, wastewater treatment and lipids for biofuel production [9].

2.3. Algal Production in Open Ponds

Compared with other cultivation methods, algal production in open ponds with low-cost construction and low operating costs has obvious advantages. Also, microalgal production in open ponds contributes to carbon emission reduction, compared with other cultivation modes. Almost all the commercial-scale microalgal products are made using this method. Common species of algae used in open ponds are shown in Table 2. To find algal species with rapid accumulation of biomass and at the same time containing large quantity of algae biofuel is a problem. Fast growing species always tend to produce less biofuel, according to Professor Zhang Yuanhui (University of Illinois at Urbana-Champaign), who used fast-growing algae to produce biocrude oil through thermochemical conversion.

TABLE 2. Biomass production of several microalgae spp. in open ponds.

Species	Yield g/(L·d)	Yield g/(m^2·d)	References
Spirulina maxima	0.21	25	[10,11]
Chlorella vulgaris	0.18		[10]
Scenedesmus obliquus	0.09		[10]
Dunaliella tertiolecta	0.12		[10]
Nannochloropsis sp.	0.09–0.31		[10]
Neochloris oleabundans	0.09		[10]
Haematococcus pluvialis		15.1–70.4	[12]
Spirulina platensis	0.06–0.18	15–51	[11,13]
Scenedesmus sp.	0.03–0.13	2.43–13.5	[14]
Chlorella sp.	0.02–2.9	1.61–25.0	[14,15]
Pleurochrysis carterae	0.02–0.22	3.20–35.2	[16]
Dunaliella salina	0.22–0.34	1.6–37.7	[17]
Spirulina sp.	0.006–0.07	2–17	[18–20]
Anabaena sp.	0.031–0.078	4.9–23.5	[14,21]
Phaeodactylum tricornutum	0.0028–0.16	2.4–11.3	[22,23]
Nannochloris sp.	0.29–0.32		[23]

2.4. Algal Culture Collections in China

For screening and selecting excellent algal strains and some species in danger of extinction from natural environments, algal culture collections are necessary to preserve the biodiversity of natural habitats, patent microalgae that adapt to practical conditions better and grow well, to protect genetic material and provide basic research resources as well. At present, several algal collection centers have been set up in China, such as the Freshwater Algae Culture Collection (FACHB) in Wuhan, which has more than 100 genera, 800 strains of freshwater algae, 3000 branches and a small amount of salt and seawater species. The Marine Biological Culture Collection Centre (MBCCC) in Qingdao has collected more than 600 strains of marine algal resources, close to them being 70% oil-producing algae species. In China there are other culture collections at some universities like Jinan University in Guangzhou, Ningbo University in Zhejiang and Ocean University of China in Qingdao. The number of cultured algal species is much less than those of described algal species and suitable media need further investigation. Microalgae isolated from wastewater treatment plant sites or real bodies of water can adapt to practical conditions well and grow better.

3. RESOURCE AVAILABILITY IN CHINA

3.1. CO_2 Sources

Generally speaking, the atmosphere contains only 0.03–0.06% CO_2. This is far from adequate to cultivate microalgae. The utilization of high concentration CO_2 from industrial flue gas for large-scale algal cultivation is a very promising method, not only for CO_2 emission mitigation but also for the production of biofuels and high value products. The CO_2 industrial point sources are mainly from power plants, cement and steel works, which account for about 91.7% of the total emissions. Power plants' relative CO_2 emissions made up 63% of the main industrial sources in 2004. The highest concentration of CO_2 was from the ammonia industry, which contained 100% CO_2 and cement production which contained 20% CO_2 (Table 3) [24].

TABLE 3. Large scale and high concentration CO_2 sources in China.

CO_2 Sources	CO_2 Relative Emissions (%)	Concentration (%)
Power plants	63	15
Cement production	19	20
Ethylene	1	12
Steel works	10	15
Refineries	3	8
Ammonia	4	100

The geographical distribution of CO_2 emissions of the main sources (power plants, cement production, steelworks, refineries, ethylene and ammonia plants) with large-scale and high concentration sources is presented in Figure 1 [24]. The distribution of CO_2 emission industrial point sources were concentrated located in Bohai Sea Bay, Yangtze River Delta, Pearl River Delta such as Shandong, Jiangsu, Hebei and Guangdong provinces.

Usually the production of per ton of algal dry biomass will assimilate about 1.83 tons of CO_2. Microalgae can use about 9% of the solar energy to produce 280 tons of dry biomass per $ha^{-1}y^{-1}$ while sequestering roughly 513 tons of CO_2 [25]. The requirements for CO_2 supply to enhance algal production and matching of CO_2 source availability with algal cultivation facilities is not just a simple question. Optimal algal growth could provide excellent opportunity for the utilization of GHG emissions. Some algae can tolerate high CO_2 concentrations and the temperature of flue gas containing trace constituents such as SO_x and NO_x. Common species of algae for CO_2utilization are shown in Table 4.

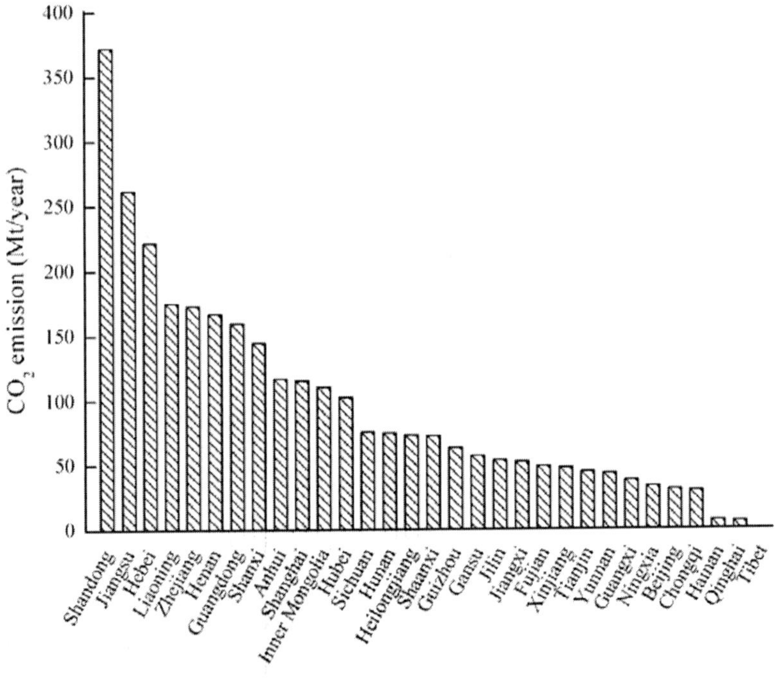

FIGURE 1. The geographical distribution of CO_2 industrial point sources.

TABLE 4. Biomass productivity, CO_2 fixation and specific growth rates for different microalgae.

Species	CO_2 (%)	Biomass Productivity (mg L^{-1} d^{-1})	CO_2 Fixation Rates (mg L^{-1} d^{-1})	μ (d^{-1})	References
Botryococcus braunii	5–10 or flue gas	26.55	496.98	0.24	[26–28]
Scenedesmus sp.	10 or flue gas	217.50			[26]
Chlorella vulgaris	5–10	104.76–310	251.64	0.29	[26,28]
Spirulina platensis	5	730	318.61	0.22	[28]
Dunaliella tertiolecta	5	420	272.4	0.21	[28]
Chlorella sp.	0.03–15	191–1484	350–13,700	0.58–0.66	[27,29,30]
Chlorococcum littorale	10–20 or flue gas	400–2500	650–17,300	1.8–1.9	[27,31–33]
Scenedesmus obliquus	6–12	40–140		0.19–0.261	[34,35]
Chlorella kessleri	6	61–90			[34]
Spirulina sp.	6–12	40–220		0.27–0.44	[35]
Aphanothece microscopica	15		5612–14,495		[36]
Euglena gracilis	5–10			1.44	[37]

3.2. Sunlight

Although the light intensity requirements of typical microalgae are relatively low compared with those of higher plants [9], the availability of abundant sunlight is equally important for the robust growth of photoautotrophic microalgae with both open and closed cultivation systems [38]. This is undoubtedly an important consideration since insolation is directly linked to yield. Different regions in China have vastly different day-length regimes (Figure 2). In the water, the light intensity is reduced with the depth and spectrum composition change. First, infrared, ultraviolet and long wavelength red light are absorbed. The short wavelength cyan and blue light are absorbed last. Algae do not need the sunlight in the same intensities (Table 5). For most algae, the appropriate range of light intensity is between 36–180 μmol m^{-2} s^{-1}, a light-dark growth is used between 12:12 and 16:8 h but some winter strains need a longer night period. Over illumination will result in photo-oxidative stress and continuous light may lead to death in some algae. Chlorophyll absorbs red and violet light, the Cyanobacteria and red algae are rich in light-harvesting pigments, which can use sunlight that chlorophyll does not absorb in other categories. The light intensity on water surface can reach thousands of μmol m^{-2} s^{-1} in sunny days, some algae suffer the limitations of light saturation photoinhibition [39]. China is a country with abundant solar energy resources, with solar radiation values between 6.5 to 7.5 kWh m^{-2} d^{-1}. The annual average illumination time of two-thirds of the area in China is more than 2000 h. The total illumination hours in Northwest and Qinghai-Tibetan Plateau are China's longest. China's Eastern, Southern and Northeast areas are the middle solar energy resources and Sichuan, Hubei and Hunan have lower solar energy resources.

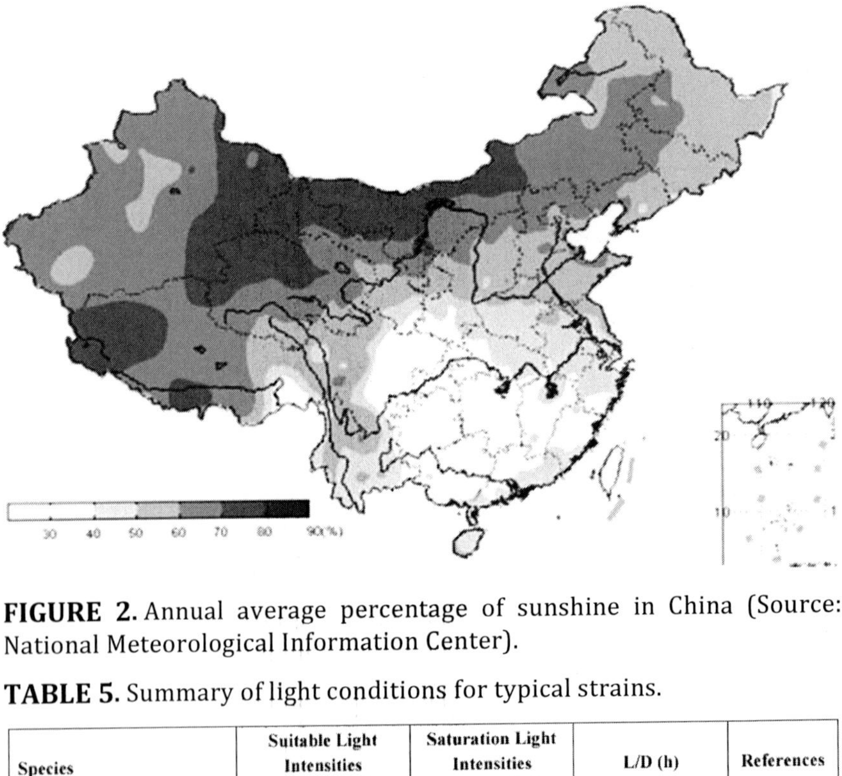

FIGURE 2. Annual average percentage of sunshine in China (Source: National Meteorological Information Center).

TABLE 5. Summary of light conditions for typical strains.

Species	Suitable Light Intensities (μmol m^{-2} s^{-1})	Saturation Light Intensities (μmol m^{-2} s^{-1})	L/D (h)	References
Spirulina platensis	10~630	~270	6:5~8:3, 24:0	[40]
Botryococcus braunii	400~1600	~800	~14:10	[41]
Nannochloropsis oculata	90~126	–	12:12	[42]
Haematococcus pluvialis	70~90	~200	~12:12	[43]
Dunaliella salina	50~90	~220	~12:12	[44]
Chlorella pyrenoidosa	120~160	~150	~10:14	[45]
Nitzschia closterium	70~160	–	~24:0, ~12:12	[46]

3.3. Water

The total water resources per capita in China are estimated at about 2245 m^3, which corresponds to one third of the World's average value and thus ranks China among the most water poor countries in the World

[47]. Freshwater sources which would already be restricted would likely not be available in large quantities for large scale algal cultivation. Nevertheless, China has large volumes of wastewater from urban households, industry and animal husbandry.

In 2008, the amount of discharge of industrial and urban household wastewater nationwide was calculated to be about 57.2 billion tons, with that of the latter reaching 33.0 billion tons [48]. From the geographical distribution of the amount of wastewater emissions in 2008 (Figure 3), we find the main wastewater distribution areas were in the southeast coast, eastern and southern China, especially, in Guangdong, Jiangsu, Shandong, Zhejiang and Guangxi. The largest amount of industrial wastewater discharge was located in Jiangsu and the largest urban household and total wastewater was presented in Guangdong. From Figure 4, we also knew that the statistics of swine wastewater emissions were steady at about 4.1–4.5 million m^3 in recent years [48]. Algae can utilize those wastewaters (Table 6), which usually contains relatively high inorganic nutrient levels and there has a great potential for the removal of nitrogen and phosphorus for algae growing and at the same time accumulating lipids in algal cells.

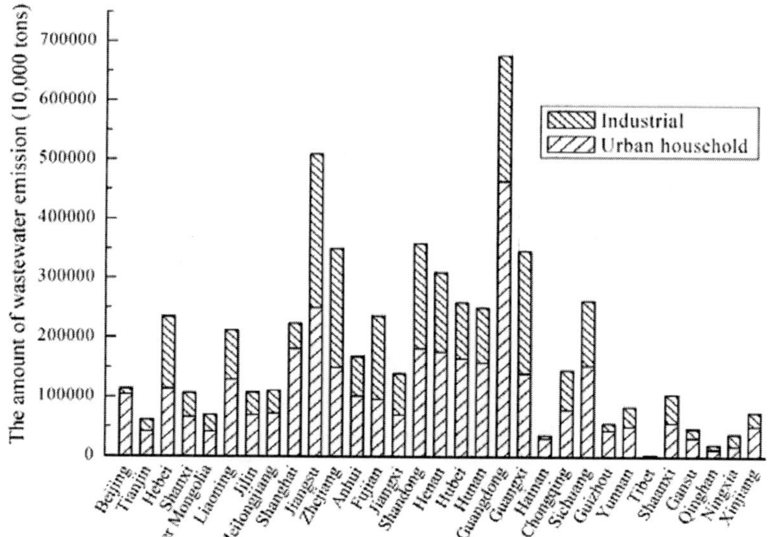

FIGURE 3. The distribution of wastewater emissions in 2008.

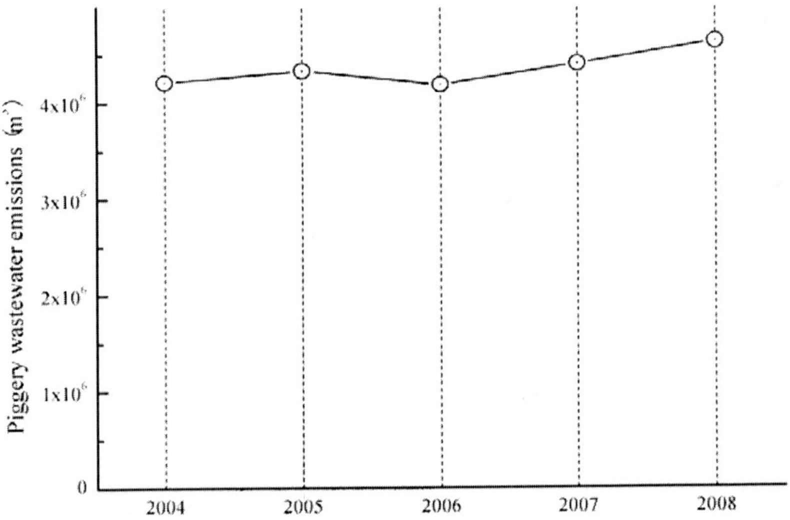

FIGURE 4. Piggery wastewater emissions in China.

TABLE 6. Typical strains cultivated in wastewater.

Species	Type of Sewage	Process	References
Scenedesmus sp.	Urban wastewater from secondary treatment or Synthetic	Batch; Semicontinuous	[49,50]
Spirulina (Arthrospira) sp.	Pig wastewater anaerobic effluents	Outdoor raceways	[51]
Chlamydomonas reinhardtii	Different stages from Metropolitan Wastewater Treatment Plant	Batch; Biocoil photobioreactor	[52]
Chlorella sp.	Dairy manure	Attached	[53]
Scenedesmus obliquus	Olive-oil mill; urban wastewater	Photobioreactor; Semicontinuous; Batch	[54–56]
Chlamydomonas globosa; *Chlorella minutissima* and *Scenedesmus bijuga*	Untreated carpet industry	Raceway ponds; Vertical tank reactors and polybags	[57]
Chlorella vulgaris	Municipal; Synthetic; Steel-making facility; Distillery	Immobilized; Photobioreactor; Batch; Microalgae pond	[56,58–62]
Chlorella sorokinia	Municipal	Immobilized	[63]
Phaeodactylum tricornutum, *Oscillatoria* sp.	Primary sewage and seawater	Corrugated raceways	[64]
Scenedesmus rubescens	Synthetic; municipal	Immobilization	[59]
Senedesmus dimorphu	Analogue wastewater	Bio-coil reactor conical flask	[65]
Monoraphidium	Tertiary treatment of urban wastewater		[66]
Aphanothece microscopica	Refinery industry	Photobioreactor	[67]
Botryococcus braunii	Livestock wastewater	Batch	[68]

3.4. Land

Land is hardly a limitation for algae cultivation. Algae cultivation uses land roughly 3.3 times more efficiently than corn, 4.3 times more efficiently than switchgrass and 5 times more efficiently than canola [69] and microalgae also do not need to compete with land used for food crops. They can encroach on pasture, grass land and even relatively sterile desert (Figure 5). The Gobi desert covering Southern Mongolia and Northern China is the highest solar radiation area, and many salt lakes are distributed there which are natural sources for nutrition of algae growth. Over the past decade, three microalgae (mainly *Chlorella* sp.) production plants have been built in Qingyang and Wuwei of Gansu Province, Erdos of Inner Mongolia Autonomous Region [4]. Those area should also be generally accessible and already serviced by roads and power.

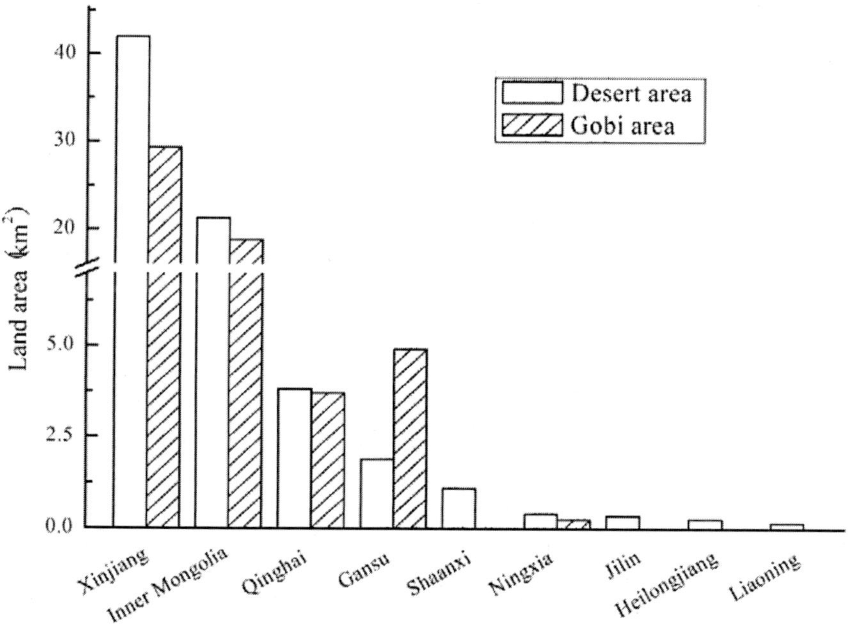

FIGURE 5. The desert and Gobi areas of China [48].

4. MICROALGAE R&D EFFORTS IN CHINA

In the 1950s, research on microalgae, *i.e.*, *Chlorella* use cultivation was started in the Institute of Hydrobiology under the auspices of the Chinese Academy of Sciences (CAS). With the passage of time in the 1980s, large scale cultivation of *Spirulina* and *Dunaliella* was listed in the national research program (the 7th Five-Year Plan). In the last decade the microalgae industry in China had been developing rapidly and has made remarkable achievements. In 1984, the Salt Research Institute, affiliated with the former China Light Industry Association, began to study *Dunaliella* cultivation for the extraction of natural carotene. In 1989, the first *Spirulina* production pilot factory was established in Chenghai Lake, Yunnan. In 1999, Guangdong Jinqiu Green Algae Co., Ltd. built a 10,000 km^2 *Chlorella* production base. In 2001, the production of algae powder was more than 10 tons. Those microalgae were all cultivated for various nutritional products [70]. According to International Energy Agency (IEA), the total world production of dry algal biomass is estimated about 10,000 tons per year, among which half of the yield is produced in mainland China.

China has started to promote its algae based energy as early as 2005. The fundamental research on energy microalgae in China has a strong scientific and technological base. Varies universities and research institutes are undertaking large numbers of national and provincial level research projects on microalgae classification, breeding and preservation technology. For example, microalgal hydrogen production at the Dalian Institute of Chemical Physics (CAS), microalgal oil production at Ocean University of China, Tsinghua University and other institutions have much more research basis. ENN Group Co., Ltd. has been developing technology to use microalgae to fix CO_2 and then to produce biofuel. Marine microalgae bioenergy projects have been established in Shenzhen, Guangdong in 2008. The main objective of the project was the cultivation of diatoms to fix CO_2 from flue gas, later on diatoms were harvested and using for biodiesel production. In 2009, CAS and China Petrochemical Corporation (Sinopec Group) decided to develop algae biodiesel technology together. It was expected that an outdoor pilot plant will be operational in about 2015 and a 10,000 ton industrial demonstration unit would be set up in the near future. However, lots of algae breeding work is still in the research and development stage and strains are not qualified for industrialization at present. The mass culturing technology of algae is still immature and lack a relevant criteria or evaluation system in China.

5. CONCLUSIONS

China has a vast sea area and a great number of rivers and lakes, resulting in abundant algal resources. For algal large-scale production, we should select construction in the vicinity of high CO_2 emission enterprises such as power plants, cement and steel works. Wastewaters from urban households, industry and animal husbandry are needed to collect and treat for algal utilization. Free sunlight and robust strain screening are important as well. In all, technology development and coordinated management for cost reduction are needed to utilize those waste and free resources. Algal bioenergy is still in its infancy and further investigation is required.

ACKNOWLEDGMENTS

This study was supported by the National Program on Key Basic Research Project of China (973 Program) (2011CB200905), the 11th and 12th Five Year Support Plan of the *Ministry of Science* and *Technology*, P. R. China (2008BADC4B00, 2011BAD14B03), the Guangdong Chinese Academy of Sciences Comprehensive Strategic Cooperation Project (2010A090100010) and the Innovation Fund for Graduate Student of China Agricultural University (15050207).

REFERENCES

1. The National Renewable Energy Laboratory (NREL). *Jet Fuel from Microalgal Lipids*; NREL: Golden, CO, USA, July 2006. Available online: http://www.nrel.gov/docs/fy06osti/40352.pdf (accessed on 16 August 2011).
2. Demirbas, A.; Fatih Demirbas, M. *Algae Energy: Algae as a New Source of Biodiesel*; Springer-Verlag: London, UK, 2010.
3. Sheehan, J.; Dunahay, T.; Benemann, J.; Roessler, P. *A Look Back at the U.S. Department of Energy's Aquatic Species Program: Biodiesel from Algae*; NREL Report No. TP-580-24190; NREL: Golden, CO, USA; July; 1998.
4. Tian, Y. *Leaving the Sunlight-Thought and Practice on the Microalgae Photoreactor*; Popular Science Press: Beijing, China, 2008.

5. Zhang, W. *China's Biodiversity: A Country Study*; China Environmental Science Press: Beijing, China, 1998.
6. Liu, R. *Checklist of Marine Biota of China Seas*; Science Press: Beijing, China, 2008.
7. Nelson, D.M.; Treguer, P. Production and dissolution of biogenic silica in the ocean: Revisedglobal estimates, comparison with regional data and relationship to biogenic sedimentation. *Global Biogeochem. Cycles* **1995**, *9*, 359–372.
8. Liu, X. *The Report of Conservation of Chinese Bio-diversity Resources*; Science Press: Beijing, China, 2004.
9. Kumar, A.; Ergas, S.; Yuan, X. Enhanced CO_2 fixation and biofuel production via microalgae: Recent developments and future directions. *Trends Biotechnol.* **2010**, *28*, 371–380.
10. Gouveia, L.; Oliveira, A.C. Microalgae as a raw material for biofuels production. *J. Ind. Microbiol. Biotechnol.* **2009**, *36*, 69–274.
11. Lee, Y. Microalgal mass culture systems and methods: Their limitation and potential. *J. Appl. Phycol.* **2001**,*13*, 307–315.
12. Huntley, M.E.; Redalje, D.G. CO_2 mitigation and renewable oil from photosynthetic microbes: A new appraisal. *Mitig. Adapt. Strateg. Global Change* **2007**, *12*, 573–608.
13. Pushparaj, B.; Pelosi, M.R.; Tredici, E. An integrated culture system for outdoor production of microalgae and cyanobacteria. *J. Appl. Phycol.* **1997**, *9*, 113–119.
14. Kanazawa, Z.; Fujita, C.; Yuhara, T.; Sasa, T. Mass culture of unicellular algae using the open pond circulation method. *J. Gen. Appl. Microbiol.* **1958**, *4*, 135–139.
15. Doucha, J.; Straka, F.; Karel, L. Utilization of flue gas for cultivation of microalgae (*Chlorella* sp.) in an outdoor open thin layer photobioreactor. *J. Appl. Phycol.* **2005**, *17*, 403–412.
16. Moheimani, N.R.; Borowitzka, M.A. The long-term culture of the coccolithophore Pleurochrysis carterae (Haptophyta) in outdoor raceway ponds. *J. Appl. Phycol.* **2006**, *18*, 703–712.
17. Garcia, G.M.; Moreno, J.; Canavate, J.P.; Anguis, V.; Prieto, A. Conditions for open-air outdoor culture of*Dunaliella salina* in southern Spain. *J. Appl. Phycol.* **2003**, *15*, 177–184.
18. Olguin, E.J.; Galicia, S.; do Merca, G.; Perez, T. Annual productivity of *Spirulina* (*Arthrospira*)and nutrient removal in a pig wastewater

recycling process under tropical condition. *J. Appl. Phycol.* *200* **2003**, *15*, 249–257.

19. Jimenez, C.; Cossio, B.R.; Labella, D.; Niell, F.X. The feasibility of industrial production of *Spirulina*(*Arthrospira*) in southern Spain. *Aquaculture* **2003**, *217*, 179–190.
20. Richmond, A. *Spirulina, Micro-Algal Biotechnology*; Cambridge University Press: Cambridge, UK, 1988; pp. 85–121.
21. Moreno, J.; Vargas, A.; Rodriguez, H.; Rivas, J.; Guerrero, M.G. Outdoor cultivation of a nitrogen-fixing marine cyanobacterium, *Anabaena* sp. ATCC 33047. *Biomol. Eng.* **2003**, *20*, 191–197.
22. Laws, E.A.; Terry, K.L.; Wickman, J.; Chalup, M. A simple algal production system designed to utilise the flashing light effect. *Biotechnol. Bioeng.* **1983**, *25*, 2319–2335.
23. Negoro, M.; Shioji, N.; Miyamoto, K.; Miura, Y. Growth of microalgae in high CO_2 gas and effects of SO_x and NO_x. *Appl. Biochem. Biotechnol.* **1991**, *28/29*, 877–886.
24. Bai, B.; Li, X.; Liu, Y.; Zhang, Y. Preliminary study on CO_2 industrial point sources and their distribution in China. *Chin. J. Rock Mech. Eng.* **2006**, *25*, 2918–2923.
25. Bilanovic, D.; Andargatchew, A.; Kroeger, T.; Shelef, G. Freshwater and marine microalgae sequestering of CO_2 at different C and N concentrations–Response surface methodology analysis. *Energy Convers. Manag.* **2009**, *50*, 262–267.
26. Yoo, C.; Jun, S.; Lee, J.; Ahn, C.; Oh, H. Selection of microalgae for lipid production under high levels carbon dioxide. *Bioresour. Technol.* **2010**, *101*, 71–74.
27. Murakami, M.; Ikenouchi, M. The biological CO_2 fixation and utilization project by rite 2. *Energy Convers. Manag.* **1997**, *38*, S493–S497.
28. Sydney, E.B.; Sturm, W.; de Carvalho, J.C. Potential carbon dioxide fixation by industrially important microalgae. *Bioresour. Technol.* **2010**, *101*, 5892–5896.
29. Fulke, A.B.; Mudliar, S.N.; Yadav, R.; Shekh, A.; Srinivasan, N.; Rishiram, R.; Krishnamurthi, K.; Devi, S.S.; Chakrabarti, T. Bio-mitigation of CO_2, calcite formation and simultaneous biodiesel precursors production using *Chlorella* sp. *Bioresour. Technol.* **2010**, *101*, 8473–8476.

30. Ryu, H.J.; Oh Kyeong, K.; Kim, Y.S. Optimization of the influential factors for the improvement of CO_2utilization efficiency and CO_2 mass transfer rate. *J. Ind. Eng. Chem.* **2009**, *15*, 471–475.
31. Kurano, N.; Ikemoto, H.; Miyashita, H.; Hasegawa, T.; Hata, H.; Miyachi, S. Fixation and utilization of carbon dioxide by microalgal photosynthesis. *Energy Convers. Manag.* **1995**, *36*, 689–692.
32. Kurano, N.; Ikemoto, H.; Miyashita, H. Fixation and utilization of carbon dioxide by microalgal photosynthesis. *Energy Convers. Manag.* **1995**, *36*, 689–692.
33. Kurano, N.; Sasaki, T.; Miyachi, S. Advances in chemical conversions for mitigating carbon dioxide. *Stud. Surf. Sci. Catal.* **1998**, *114*, 55–63.
34. Morais, M.G.; Costa, J.A.V. Isolation and selection ofmicroalgae fromcoal fired thermoelectric power plant for biofixation of carbon dioxide. *Energy Convers. Manag.* **2007**, *48*, 2169–2173.
35. De Morais, M.G.; Costa, J.A. Biofixation of carbon dioxide by *Spirulina* sp. and *Scenedesmus obliquus*cultivated in a three-stage serial tubular photobioreactor. *J. Biotechnol.* **2007**, *129*, 439–445.
36. Jacob-Lopesa, E.; Revahb, S.; Hernández, S. Development of operational strategies to remove carbon dioxide in photobioreactors. *Chem. Eng. J.* **2009**, *135*, 120–126.
37. Chae, S.R.; Hwang, E.J.; Shin, H.S. Single cell protein production of *Euglena gracilis* and carbon dioxide fixation in an innovative photo-bioreactor. *Bioresour. Technol. 20* **2006**, *97*, 322–329.
38. *National Algal Biofuels Technology Roadmap*; U.S. Department of Energy: Washington, DC, USA, 2010.
39. Chen, F. *Microalgae Biotechnology*; China Light Industry Press: Beijing, China, 1999.
40. Wang, B.; Mao, H.; Jiao, Z.; Zheng, Y. The analysis of cultivating *Spirulina* on agro-meteorological conditions in chenghai lake. *Yunnan Geogr. Environ. Res.* **2009**, *6*, 25–28.
41. Yin, D.; Geng, Y.; Mei, H. The effects of several environmental factors on the photosynthesis of *Botryococcus braunii*. *J. Wuhan Bot. Res.* **2008**, *26*, 64–69.
42. Jiang, X. Effects of temperatures, light intensities and nitrogen concentrations on the growth and fatty acid compositions of *Nannochloropsis oculata*. *Mar. Sci.* **2002**, *8*, 9–13.
43. Lu, K.; Jiang, X.; Zhai, X. Effects of illumination on growth of *Haematococcus pluvialis*. *Hebei Fish.* **2002**, *6*, 6–8.

44. Lin, G. Study on the conditions of high density cultivation of *Dunaliella salina*. *Shandong Fish.* **2007**, *9*, 41–44.
45. Wen, G.; Liang, W.; Li, Z. Studies on the ecological factors of four microalgae populations in mixed-culture.*Prog. Fish. Sci.* **2009**, *30*, 142–148.
46. Shi, J.; Pan, K. Effects of different light intensities on growth and biochemical composition of *Nitzschia closterium* f. minutissima and *Isochrysis galbana* Parke 8701.*J. Fish. Sci. China* **2004**, *11*, 121–128.
47. Li, W.; Wang, R. Water Resources Ecosystem Management in China-Challenges and Opportunities. In Proceedings of the Ecosystem Service and Sustainable Watershed Management in North China International Conference,, Beijing, China, 23–25 August 2000.
48. National Bureau of Statistics of China. *China Statistical Yearbook*; China Statistics Press: Beijing, China, 2004–2009.
49. Li, X.; Hu, H.; Yang, J. Lipid accumulation and nutrient removal properties of a newly isolated freshwater microalga, *Scenedesmus* sp. LX1, growing in secondary effluent. *New Biotechnol.* **2010**, *27*, 59–63.
50. Voltolina, D.; Gómez-villa, H.; Correa, G. Biomass production and nutrient removal in semicontinuous cultures of *Scenedesmus* sp. (chlorophyceae) in artificial wastewater, under a simulated day-night cycle. *Vie Milieu* **2004**, *54*, 21–25.
51. Olguín, E.J. Annual productivity of *Spirulina* (*Arthrospira*) and nutrient removal in a pig wastewater recycling process under tropical conditions.*J. Appl. Phycol.* **2003**, *15*, 249–257.
52. Kong, Q.; Li, L. Culture of Microalgae *Chlamydomonas reinhardtii* in wastewater for biomass feedstock production.*Appl. Biochem. Biotechnol.* **2010**, *160*, 9–18.
53. Michael, B.; Wen, Z. Development of an attached microalgal growth system for biofuel production.*Appl. Microbiol. Biotechnol.* **2010**, *5*, 525–534.
54. Hodaifa, G. Daily doses of light in relation to the growth of *Scenedesmus obliquus* in diluted three-phase olivemill wastewater.*J. Chem.Technol. Biotechnol.* **2009**, *84*, 1550–1558.
55. Voltolina, D. Nitrogen removal and recycling by Scenedesmus obliquus in semicontinuous cultures using artificial wastewater and a simulated light and temperature cycle. *Bioresour. Technol.* **2005**, *96*, 359–362.

56. Martinez, M.E.; Sancheze, S.; Jimenez, J.M. Nitrogen and phosphorus removal from urban wastewater by the microalga *Scenedesmus obliquus*. *Bioresour. Technol.* **2000**, *73*, 263–272.
57. Chinnasamy, S.; Bhatnagar, A.; Claxton, R. Biomass and bioenergy production potential of microalgae consortium in open and closed bioreactors using untreated carpet industry effluent as growth medium.*Bioresour. Technol.* **2010**, *101*, 6751–6760.
58. De-Bashana, L.E.; Moreno, M.; Hernandez, J.P. Removal of ammonium and phosphorus ions from synthetic wastewater by the microalgae *Chlorella vulgaris* coimmobilized in alginate beads with the microalgae growth-promoting bacterium *Azospirillum brasilense*. *Water Res.* **2002**, *36*, 2941–2948.
59. Shi, J. Removal of nitrogen and phosphorus from wastewater using microalgae immobilized on twin layers: An experimental study. *J. Appl. Phycol.* **2007**, *19*, 417–423.
60. Kapdan, I. K.; Aslan, S. Application of the Stover-Kincannon kinetic model to nitrogen removal by *Chlorella vulgaris* in a continuously operated immobilized photobioreactor system. *J. Chem. Technol. Biotechnol.* **2008**,*83*, 998–1005.
61. Yun, Y.; Sun, B. Carbon dioxide fixation by algal cultivation using wastewater nutrients. *J. Chem. Technol. Biotechnol.* **1997**, *69*, 451–455.
62. Travieso, L.; Benítez, F.; Sánchez, E. Performance of a laboratory-scale microalgae pond for secondary treatment of distillerywastewaters. *Chem. Biochem. Eng.* **2008**, *22*, 467–473.
63. De-Bashan, L.E.; Hernandez, J.-P.; Morey, T. Microalgae growth-promoting bacteria as "helpers" for microalgae: A novel approach for removing ammonium and phosphorus from municipal wastewater.*Water Res.* **2004**, *38*, 466–474.
64. Rupert, J.; Paul, J.; Valerie Smith, J. Wastewater nutrient removal by marine microalgae grown on a corrugated raceway. *Water Res.* **1997**, *31*, 1701–1707.
65. Liu, Y.; Ruan, R.; Kong, Q. Mass culture of high oil content microalgae on wastewater and power plant flue gases. *Chin. J. Bioprocess Eng.* **2008**, *3*, 29–33.
66. Larsdotter, K.; Jansen, J. Phosphorus removal from wastewater by microalgae in Sweden a year round perspective. *Environ. Technol.* **2010**, *31*, 117–123.

67. Jacob-Lopes, E. Biotransformations of carbon dioxide in photobioreactors. *Energy Convers. Manag.* **2010**, *51*, 894–900.
68. Shen, Y.; Yuan, W.; Pei, Z.; Mao, E. Culture of microalga *Botryococcus* in livestock wastewater. *Trans. ASABE* **2008**, *51*, 1395–1400
69. Clarens, A.F.; Resurreccion, E.P.; White, M.A; Colosi, M.; White, L.M. Environmental life cycle comparison of algae to other bioenergy feedstocks. *Environ. Sci. Technol.* **2010**, *44*, 1813–1819.
70. Ling, F. The trend of microalgae industry in China. *Ecol. Econ.* **2004**, *2*, 58–61.

Index